Ageism and Mistreatment of Older Workers

Patricia Brownell • James J. Kelly
Editors

Ageism and Mistreatment of Older Workers

Current Reality, Future Solutions

Springer

Editors
Patricia Brownell
Graduate School Social Service
Fordhan University
New York, NY, USA

James J. Kelly
Office of the Provost and Executive Vice
Menlo College
Atherton, CA, USA

ISBN 978-94-007-5520-8 ISBN 978-94-007-5521-5 (eBook)
DOI 10.1007/978-94-007-5521-5
Springer Dordrecht Heidelberg New York London

Library of Congress Control Number: 2012951047

Printed on acid-free paper

Springer is part of Springer Science+Business Media (www.springer.com)

We dedicate this book to Dr. Mary Ann Quaranta, who was our inspiration, our mentor, and our role model throughout the years we were privileged to know her and work with her. She served as Director of National Catholic Charities, Dean of the Fordham University Graduate School of Social Service, and Provost of Marymount College. She was a strong leader and a visionary, and embodied active and productive aging. When the time came, she stared ageism in the face and never blinked.

Foreword

Over 40 years ago Robert Butler (1969) coined the term *ageism*. From the start, and before the term was made famous in his Pulitzer prize-winning book, *Why Survive? Growing Old in America* (Butler 1976), it struck a responsive chord among those of us who were students, researchers, educators, or practitioners in the then new and emerging field of gerontology. We recognized that this new term captured what we were witnessing in the way many older people were marginalized and discriminated against in health care (Kane and Kane 2005), housing, and other services.

For me, ageism was epitomized by a story told to me by the widow of a prominent Canadian. She confided, "I shall never forget the look on my husband's face when, several years after he retired, a well-meaning but thoughtless individual approached him at a cocktail party and said, "I know you, you *used to be* Louis Rasminsky, Governor of the Bank of Canada. Your signature was on all our money." I remember thinking at the time, "To be sure, we leave behind some roles as we move through life from infant to child, youth, young, middle aged and older adult. But still, we are *persons*—with a present and a future as well as a past!"

Although abuse and its underpinning, ageism, occur in many settings, this book focuses on the workplace. Exploring ageist and abusive attitudes and behaviors in the workplace is timely because 2011 is the year that the leading edge of the Baby Boom generation turns age 65. Statistics from the USA (Munnell 2011) and Canada (Carriere and Galarneau 2011) already show that fewer workers are taking early retirement. We can expect to see this trend continue for the foreseeable future, as more and more persons aged 65 and over stay in the paid labor force, in some cases by choice, and in others, out of economic necessity.

The editors and authors of this book come from a variety of backgrounds and professions—social work, law, business, sociology, and education. Their work backgrounds are in settings ranging from academia to human resources departments to adult protection services. A number of them are members of my organization, the International Network for Prevention of Elder Abuse (INPEA). In 2008, INPEA organized an international conference entitled *Ageism, Aging and Abuse: Moving from Awareness to Action*. An edited collection of presentations from the conference was published 2 years later (Gutman and Spencer 2010). The current book, edited by Patricia Brownell and James J. Kelly, builds on and expands issues and themes contained in the earlier publication and introduces new ones. Robert Butler would be pleased by the breadth and depth of ideas expressed in this book. He, and Alex Kalache, former director of the World Health Organization's Life Course and Ageing Programme, would also be pleased to see the way in which the concept of *Active Ageing* (World Health Organization 2002) is promoted in the book—that is, the idea that older people can remain active, economically productive members of society if ageist barriers to their participation are recognized and dismantled.

In fostering active aging it is vital to remember the insurance principle: not every older person gets sick or dies at the same time. It is also important to recognize that the elderly are not a homogeneous group in terms of health and functional status. Although at any given time, 5% are sufficiently debilitated that they require care in a skilled nursing facility and another 10% live at home with major restrictions on their mobility, the remaining 85% can do most of what they want to do most of the time. Thus, they continue to be a valuable resource for their family, community, and country.

This book focuses on North America, but population aging is taking place around the globe. Sadly, ageism, elder abuse and neglect and, more generally, violation of the human rights of older persons are also global phenomena. This was recognized in 2002 by the United Nations (UN) and the nations that signed the *Madrid International Plan of Action on Ageing* (See Priority C, Issue 3, United Nations, 2003, p.43). Now, 10 years later, the UN's Madrid+10 review is underway, and progress toward achieving the goals of the *Plan* is being evaluated. This is a propitious time for the voices of seniors and their advocates to be heard in support of a binding human rights instrument, such as the Convention for the Human Rights of Older Persons, proposed by the Global Alliance for the Rights of Older Persons, a collaboration of nine non-governmental organizations dedicated to improving the health and well-being of older persons.

Chapter 10 in this book discusses the Convention and its implication for older persons' rights in the workplace. I encourage readers to think broadly about what its implementation would mean for the wage earning senior of today and tomorrow, and for the persons their work supports. These could include spouses, children, grandchildren, and even great-grandchildren from current and former marriages. In these times of changing demographics and economics, we need to think creatively about how to support and ensure the quality of life of frail elders and harness the resources, skills, and energy of those able and willing to continue in the paid labor force.

International Network for Prevention
of Elder Abuse (INPEA)
October, 2011.

Gloria M. Gutman, PhD, OBC

References

Butler, R. (1969). Ageism: Another form of bigotry. *The Gerontologist, 9*(4, Pt.1), 243–246.

Butler, R. (1976). *Why survive? Growing old in America*. New York: Harper & Rowe.

Carriere, Y., & Galarneau, D. (2011). Delayed retirement: A new trend? In: *Perspectives on labour and income*, Statistics Canada, Catalogue no. 75-001-X. Ottawa: Statistics Canada.

Gutman, G., & Spencer, C. (2010) (Eds.) *Aging, ageism and abuse – Moving from awareness to action*. London: Elsevier Insights.

Kane, R. L., & Kane, R. A. (2005). Ageism in healthcare and long-term care. *Generations*, 29(3), 49–54.

Munnell, A. H. (2011). What is the average retirement age? *Center for Retirement Research at Boston College Bulletin*, August 2011, No. 11–11.

United Nations. (2003). *Political declaration and Madrid international plan of action*. Available from http://social.un.org/index/Ageing/Resources/MadridInternationalPlanofActiononAgeing.aspx

World Health Organization. (2002). *Active ageing – A policy framework*. Available from www.who.int/ageing

Preface

Ageism is a term defined by Dr. Robert Butler and others, who have examined this phenomenon as a factor limiting options for older adults in society. It was coined by Dr. Robert Butler in 1968 to describe the systematic stereotyping of and discrimination against older people because they are old. The purpose of this book is to promote an understanding of ageism in the workplace as a social problem that merits attention by researchers, policy makers, and practitioners.

The impact of ageism has been examined in relation to family and societal abuse and exclusion; however, it has not been examined in any depth in relation to employment and civic engagement. This is in spite of the promotion by governments and civil society of older adults as viable contributors to society as workers and volunteers in work settings.

An Aging Population

Although the authors in this edited volume primarily address ageism in the workplace in the United States and Canada, it is by no means a problem specific to North America. Older adults now make up a larger proportion of the world's population than ever before. The twentieth century experienced a revolution in longevity, and, according to the United Nations (UN) Population Division, this remarkable demographic transition will result in the old and young representing an equal share of the world's population by 2050. This population dynamic has been recognized by international organizations, such as the UN, as well as by national non-governmental organizations such as AARP (formerly the American Association for Retired Persons).

However, the professional literature has until recently said little about how this "age-quake" or "silver tsunami" will transform society and the global workforce. The editors of this book believe that workers, workforce managers, work environments, and organizational cultures have not yet accommodated to these transformative socio-demographic changes. As a result, intentional and unintentional ageist practices pervade the contemporary workforce. The aging of the workforce has generated a growing body of policy responses and these require a new understanding of the roles of older adults in the workforce. Employers and managers will also have to devise new remedies to change ageist values and prevent mistreatment of older workers. This book provides the necessary policy and practice knowledge to assist employers, managers, researchers, policy makers, and other professionals from multi-disciplinary perspectives to understand the implications of an older workforce within a human rights context.

Securing a Place in the Workforce for Older Workers

Older adults, particularly in developed western countries like the United States and Canada, are living not only longer, but healthier lives, and many seek to remain productive in the workplace, through both paid employment and volunteer work. Their doing so necessitates an agenda that calls for greater older adult participation in political, legislative, and social policy spheres. Along with this will come empowerment of older adults and their full integration into the workplace, so they might share in its benefits.

These trends are not confined to North America. A comprehensive international aging agenda and plan of action for its implementation is included in the International Plan of Action and Political Declaration adopted in Madrid, Spain, in 2002. This international aging agenda, signed onto by most of the 193 member nations of the UN, emphasizes the importance of older people and the vital role they play in society, including the workplace.

Despite these developments, images of aging continue to be perpetuated that portray older adults as burdens, a drain on society, and incapable of keeping up with the demands of the modern workplace. These images are common in contemporary workplaces. As a result, many workplace environments remain inhospitable to older adult workers. This places older adults at risk of discrimination and mistreatment in the workplace. Barriers to promotion of economic, health, and social well-being of older adults exist through discriminatory workplace policies and practices targeting older adult workers who may be the focus of ageist stereotypes and assumptions held by administrators and co-workers. Discriminatory policies affecting mandatory retirement based on age still exist in workplace settings ranging from academia to the UN. Other workplace policies and practices are more subtle, but nonetheless discriminatory and harmful to older adults in the workplace.

Employee mistreatment in the workforce has been a topic of interest in Europe, Australia, and Canada in the last 10 years, and in gender studies in the United States. However, to date there has been little discussion, research, or policy development to ensure protection and well-being of older adult workers in a workforce that has recently been challenged by recession, global financial instability, and technological change. This book seeks to remedy these knowledge gaps.

Overview of the Book

Ageism and Mistreatment of Older Workers: Current Reality, Future Solutions is designed for gerontologists, researchers, policy makers, and practitioners who are committed to improve the quality of life of older people. The goal is to allow older people to live with vitality, dignity, and rights, as active and productive members of society. The editors, an emerita professor of social work and international expert in elder abuse, and a business school president and prominent gerontologist, have teamed up to compile a comprehensive book on research, policy, and practice issues in workplace ageism and mistreatment of older adults. This book also proposes remedies to address these problems, as well as directions for future research, policy, and practice.

This book brings together chapters that explore the challenges and opportunities of an aging workforce as seen by experts in the fields of human rights, gerontology, social work, counseling, law, human resources, and vocational psychology. Although there have been several recent books on aging and work, ageism, workforce bullying, and elder mistreatment, none has combined all these issues within a human rights framework and focused on research, policy, and practice issues specific to ageism and older adult mistreatment in the workplace.

The book's ten chapters identify problems older adults face in the workplace due to discrimination, ageism, and mistreatment. These are phenomena—widely studied among women and minorities—that researchers have shown lead to post-traumatic stress, poor health and mental health outcomes, and premature disengagement from work and civic roles such as volunteerism.

The chapters are grouped into three sections: A Human Rights Perspective and Conceptual Framework (Chaps. 1 and 2); Current Reality (Chaps. 3, 4, and 5), and Future Solutions (Chaps. 6, 7, 8, 9, and 10).

Chapters 1 and 2 provide conceptual frameworks for understanding and responding to ageism and older adult mistreatment in the workplace. In Chapter 1, Denise Gosselin Caldera provides a human rights framework for the book. She focuses on one particular international aging agenda that has been endorsed by most countries of the world under the auspices of the UN: the Madrid International Plan of Action on Ageing (MIPAA) 2002. In particular, Caldera focuses on MIPAA Priority Areas 1: *Older Persons and Development*; and 3: *Promoting Enabling Environments*. This chapter sets the stage for subsequent chapters as relevant to the 2012 ten-year review of MIPAA, along with the history of emerging policies on mainstreaming older adults in society, with particular relevance to North America. Caldera ends her chapter with a brief discussion of the international movement toward a UN Convention for the Human Rights of Older Persons and its implications for protecting the rights of older workers.

A theoretical and conceptual framework for understanding ageism and older adult mistreatment in the workplace is presented in Chap. 2. Authors Patricia Brownell and Mebane Powell argue for the inclusion of older adults in an expanded definition and framework to be used by researchers, policy makers, and practitioners in the study of worker mistreatment. Brownell and Powell suggest that gerontologists need to embrace workplace *ageism* and *abuse* as co-occurring, rather than separate, issues. Only by doing so, will employers and managers implement effective policies to combat mistreatment of older workers. At the same time, elder abuse researchers and practitioners need to move beyond narrow conceptualizations of abuse and mistreatment of vulnerable older adults to include mistreatment of older workers who are targeted because of their age.

Chapters 3 through 5 examine workplace realities as they affect older adult workers in both the workplace and retirement. In Chap. 3, Amy Blackstone reviews research findings on older adult mistreatment in the workplace, focusing on harassment as a form of mistreatment from the perspective of older adult workers themselves. There have been few empirical studies on harassment and abuse of older adults in the workplace. Blackstone reviews and summarizes research on this topic to date, with a special focus on a recent study funded by the National Science Foundation on harassment of older adults in the workplace. Both qualitative and quantitative measures are presented and discussed, providing a foundation for the subsequent chapters. Blackstone also highlights results from her recent study of harassment of older workers, arguing that the power older people hold at work, at home, and in their communities shapes their workplace harassment experiences. Based on the findings of this study and a review of the literature, Blackstone concludes that workplace isolation is one of the most common harassment experiences among older workers. She also posits that gender differences are less prominent than anticipated, and that many older workers remain silent about their harassment experiences. She concludes by considering what these findings suggest about improving older people's workplace experiences.

In Chap. 4, Mebane Powell presents the results of a qualitative study of United States newspaper articles about older workers and retirement, undertaken as a means of gauging public perceptions of older workers, as well as ageism and abuse in the workplace. The recent economic crisis in the United States has led to an increase in media coverage of older workers who are laid off, or forced to retire or work past their planned retirement age. Embedded in the reports are the undertones of ageism and abuse of older workers in the workplace. Powell identifies trends in perceptions of older workers as portrayed in newspapers and highlights the lack of resources for supporting older adults who seek work or struggle to remain in the workforce. She concludes that ageism and mistreatment of older people in the workplace serve as ways to deny basic human rights to older individuals through the use of negative stereotypes and economic oppression. On a more optimistic note, Powell also discusses newspapers' role in promoting positive images of older adult workers and focusing attention on the successful efforts of some businesses to develop and maintain age-friendly work environments.

Chapter 5 presents Lynn McDonald's historical examination of ageist attitudes and beliefs embedded in employment and retirement policies and practices relevant to older persons. Repeal of mandatory retirement laws and new policies to attract and retain older workers raise questions that McDonald explores in this chapter: Will older people be forced to toil longer to stay financially healthy? Will they change careers later in life to explore new interests? How will working longer affect their health? How much will older workers affect the ambitions and working styles of younger colleagues? Will companies have to change their health and benefit plans to accommodate older people? McDonald discusses implications for both individuals and companies of employing workers beyond the mandatory retirement age. These include issues such as unequal power relationships between older workers and their younger colleagues and managers. She also addresses issues related to international political economy.

Chapters 6 through 10 identify solutions to ageism and mistreatment of older adults from organizational, counseling, and legal perspectives. In Chap. 6, Elizabeth Ciampa and Roslyn Chernesky discuss ageism in the workplace from an organizational perspective. They review the literature on older people and the workforce, and discuss ways in which workplace environments perpetuate or mitigate ageist workplace practices. Ciampa and Chernesky argue that administrators and managers in non-profit and for-profit service and business workplace environments seek to ensure the best services for clients, patients, and customers. However, the growing number of older people in the workforce challenges ageist workforce assumptions and has the potential for creating conflict and tension in the workplace. Gender, racial, and ethnic diversity compound the challenges for administrators and workers alike. In light of these challenges, Ciampa and Chernesky discuss practice strategies for administrators and managers to maintain age-friendly and abuse-free work environments.

James Woolever's Chap. 7 builds on the discussion in Chap. 6 by examining in depth the influence of human resource departments in addressing ageism, discrimination, and harassment of older workers. Businesses, health and mental health institutions, and social service agencies all include older adult employees and managers. Especially during economic downturns, when there is pressure on management to cut costs by downsizing, older workers may be targets for dismissal: they may earn higher salaries due to longevity in the job and be perceived as risks for higher health care costs. Woolever highlights insidious and destructive strategies sometimes advised by attorneys as means for terminating older workers. One such strategy is an abrupt termination based on a "reorganization" in which older workers are laid-off and told they are ineligible to apply for any position in the newly reorganized organization. Another termination strategy is a type of authorized bullying. In this strategy, employers arrange for co-workers to complain about the targeted worker, advise the older worker his performance is slipping, write up so-called infractions, and issue a series of warnings to the worker, culminating in dismissal. These strategies can result in unnecessary and long-lasting emotional and physical trauma to the older person, and they serve to perpetuate an ageist workplace environment.

Counter-balancing these trends, anti-discrimination laws prevent some targeting based on age alone, but leave much discretion to employers as to circumstances under which it is permissible to terminate older workers. In addition, the International Labour Organization (ILO)—a tripartite UN organization with government, employer, and worker representation intended to promote social justice and internationally recognized human and labor rights—has promoted age-friendly policies such as buy-outs, phase-out packages, and other incentivized self-selected layoff strategies.

Counseling interventions for older workers who are victims of discrimination and mistreatment in the workplace are discussed by Margo A. Jackson in Chap. 8. Jackson argues that internalized ageism, depression, and uncertainty about competency may make older victims of workplace mistreatment reluctant to protect themselves from unfair treatment, co-worker harassment, and bullying. Thus, they may fail to file complaints or pursue justified legal recourse. Although Employee Assistance Programs (EAPs) may be available, older adult victims of workplace discrimination and abuse may be fearful of utilizing them. Often, psychologists, counselors, psychiatric nurses, and social workers who practice with victims of abuse, including workplace abuse, are not knowledgeable about the specific issues that confront older workplace abuse victims. In this chapter, Jackson provides theoretical and practical information to practitioners with older clients who suffer workplace mistreatment.

In Chap. 9, Charmaine Spencer describes current and pending legislation to protect older workers from discrimination and workplace abuse primarily but not exclusively in North America and Europe. In her review of anti-age discrimination legislation in North America, she concludes that US legislation at the federal level offers limited protections for older workers. Much the same can be said for Canadian laws at the federal and provincial level. Even so, Spencer notes that available remedies currently are being tested in US courts at an accelerated rate due to the current economic downturn. Spencer places these laws in the broader context of how anti-bullying and ageism in the workplace are being framed, and the intersection of ageism with other types of discrimination, as well as older workers' ability to access and make use of legal remedies. Until progress is made in these three areas, Spencer argues, older workers' situations will continue to be precarious.

In recent years, there has been a movement in the United States to pass state laws to prevent workplace bullying and mistreatment, resulting in a number of pending 'healthy workplace' bills in almost all states. In Canada, although older workers' precarious position is sometimes recognized, the need for additional protections in the workplace has received little, if any, attention to date. In contrast, the European Union has passed into law stringent anti-bullying workplace policies according to the International Labour Organization that—in combination with anti-age discrimination laws—offer more protection to older workers than legislation passed in North America to date.

Israel Doron, Bethany Brown, and Susan Somers devote Chap. 10 to a discussion of ageism in the workplace and the movement toward a UN Convention for the Human Rights of Older People. When the UN Declaration of Human Rights was passed in 1948, it included few references to older adults as a distinct category of people in need of protection. In this chapter, Doron, Brown, and Somers review a recent effort to correct this omission: the international movement toward a Convention for the Human Rights of Older People. The Madrid 2002 International Plan of Action in Ageing addresses discrimination against older adults as part of the Second World Assembly on Ageing held in Madrid, Spain, in April 2002. The authors conclude that a UN Convention for the Human Rights of Older People would raise the issue of ageism generally, and in the workplace specifically, as a human rights issue.

Doron, Brown, and Somers place ageism and abuse in the workplace within the current movement to extend human rights protections to older adults. This movement emphasizes social inclusion and the right of older adults to remain productive and active in the workplace with dignity and freedom from abuse. In this chapter, the authors give gerontologists, lawyers, and other interested practitioners in North America a better understanding of the global picture of ageism in the workplace as it relates to international and regional human rights agreements. The legal landscape regarding age discrimination is very different around the world. In many undeveloped countries age discrimination is unrecognized under local laws. Even in many developed countries, age-based employment policies, such as mandatory retirement, a maximum age limit for recruitment, and age-based workforce layoffs are common and legal practices.

Currently, the legal field of human rights for older persons is in a state of flux. New laws, court rulings, and international instruments and policies are being established. In 2011, an ongoing working group under the auspices of the UN began to examine older people's human rights in nations around the world. The purpose of the UN working group is to better understand gaps in social protection of older adults under current national and international law, and determine if stronger international human rights instruments are warranted.

This book provides a comprehensive view of ageism in the workplace and how it leads to, and reflects, various forms of older worker mistreatment. It proposes a human rights framework for understanding, and for intervening, to alleviate the workplace mistreatment of older adults. The editors of this collection of chapters hope that the knowledge assembled here will be used by gerontologists, researchers, policy makers, helping professionals, employers, managers, and others to create just working conditions and secure environments for older workers. The ultimate beneficiary will be society at-large.

New York, New York — Patricia Brownell
Atherton, California — James J. Kelly

Editor Biographies

Patricia Brownell, Ph.D., LMSW, is Associate Professor Emerita of Social Service at Fordham University. Dr. Brownell represents the International Network for the Prevention of Elder Abuse (INPEA) on the CoNGO Committee on Ageing, United Nations/New York, and serves as the Chair of the Sub-Committee on Older Women (SCOW). Dr. Brownell also serves on the Executive Committee of the National Committee for the Prevention of Elder Abuse (NCPEA) and chairs its Public Policy Committee. She was on the faculty of the Fordham University Graduate School of Social Service for 20 years and is John A. Hartford Geriatric Social Work Faculty Scholar. She was Faculty Director of the Institute for Women and Girls at Fordham University until December 2011 and is a New York Academy of Medicine Social Work Fellow, a Gerontological Society of America Fellow, and Past President of the State Society on Aging of New York. She currently serves on the New York Academy of Medicine Age Friendly Cities Commission, Sub-Commission on Age Friendly Universities and Colleges, and teaches elder abuse at Molloy College.

James J. Kelly, Ph.D., LCSW, is President of Menlo College and Immediate Past President of the National Association of Social Workers. With nearly 150,000 members, NASW is the largest professional organization of social workers worldwide. Dr. Kelly is a tenured social work educator, having worked for 17 years in the California State University system. Most recently he was Provost and Professor at Menlo College, and Associate Vice President of Continuing and International Education and Interim Provost at California State University, East Bay. A former consultant to the United Nations on aging, he has published and presented widely in the field of gerontology. He has extensive global experience in higher education administration including supervision of MBA programs in Austria, Russia, Singapore, and Hong Kong, and a CPA program in Japan. He was Post-Doctoral Clinical Fellow in Psychiatry at the Sepulveda Veterans Administration Medical Center, UCLA School of Medicine, Department of Psychiatry, and is a Gerontological Society of America Fellow.

Contents

Contributors

Amy Blackstone Ph.D. Department of Sociology, University of Maine, Orono, Maine

Bethany Brown, J.D. HelpAge International, London, UK

Denise Gosselin Caldera, M.A. International Network for the Prevention of Elder Abuse, New York, NY, USA

Roslyn Chernesky, D.S.W Department of Social Service, Fordham University, New York, NY, USA

Elizabeth Ciampa, M.A. Fordham University, New York, NY, USA

Israel Doron, J.D. Department of Gerontology, University of Haifa, Haifa, Israel

Gloria M. Gutman, Ph.D. INPEA, Simon Fraser University, Vancouver, BC, Canada

Margo A. Jackson, Ph.D. Division of Psychology and Educational Psychology, Graduate School of Education, Fordham University, New York, NY, USA

Lynn McDonald, Ph.D. Faculty of Social Work, Institute for Life Course and Aging, University of Toronto, Toronto, ON, Canada

Mebane Powell, M.S.W. Graduate School of Social Service, Fordham University, New York, NY, USA

NYC Department for the Aging, Office of Research, New York, NY, USA

Susan B. Somers, J.D. International Network for the Prevention of Elder Abuse, Vancouver, Canada

Charmaine Spencer, J.D. Gerontology Research Centre, Simon Fraser University, Vancouver, BC, Canada

Canadian Network for the Prevention of Elder Abuse, Toronto, ON, Canada

James Woolever, Ph.D. Professional Studies Program, Menlo College, Atherton, CA, USA

Part I
Understanding Mistreatment of Older Workers: A Human Rights Perspective and Conceptual Framework

Chapter 1
Older Workers and Human Rights: National and International Policies and Realities

Denise Gosselin Caldera

Age discrimination continues to be a challenge for older persons. Stereotypical narratives of older workers, bias against an older person's capabilities, policies that pit younger workers against older workers, and abusive and discriminatory practices by employers discourage older workers, Nevertheless, governments, employers, and older persons themselves are recognizing the social and economic impact that age discrimination is having in the workplace. Cognizant of the economic implications of ageism, governments in Canada and the United States have enacted legislation that enforces existing policies or develop new ones to protect older workers (for further discussion see Chapter 9). Some employers, acknowledging the skills and talent lost when older employees retire, are rehiring their older staff members to mentor your workers. And older persons themselves, informed of their rights, (United Nations 2009) are fighting back and forging intergenerational alliances with young people to ensure that development policies are inclusive of all age groups (Gray Panthers Action Network 2012).

The right to work is one of the most fundamental human rights. Inherent in this right is freedom from discrimination and arbitrary dismissal (Office of the High Commissioner for Human Rights 2003). However, for millions of older persons around the world, human rights are denied or limited simply because of age. As the population of older persons increases in all parts of the world, public and private sectors, and older persons themselves, need to proactively combat ageist practices and negative stereotyping of older persons. According to the Office of the High Commission for Human Rights, it is the duty of the States to promote and protect all human rights and fundamental freedoms, regardless of their political, economic, and cultural systems. All human rights are indivisible, whether they are civil and political rights, such as the right to life, equality before the law and freedoms of expression; economic, social and cultural rights, such as the right to work, social

D.G. Caldera, M.A. (✉)
International Network for the Prevention of Elder Abuse, New York, USA
e-mail: dgosscaldera@gmail.com

P. Brownell and J.J. Kelly (eds.), *Ageism and Mistreatment of Older Workers: Current Reality, Future Solutions*, DOI 10.1007/978-94-007-5521-5_1,

security and education or collective rights, such as the rights to development and self-determination, are indivisible, interrelated and interdependent. The improvement of one right facilitates advancement of the others. Similarly, the deprivation of one right adversely affects the others (Office of the High Commission for Human Rights 2003).

The Nature of United Nations Decisions

United Nations (UN) negotiations take place within forums that may appear to be independently organized but, in fact, are closely linked. All decisions do not carry the same weight, however, and their importance depends on what kind of document they appear in and which body issues them (Sidhu 2007). The following discussion summarizes the nomenclature for various levels of UN decisions, in order from least to most binding on member states.

UN *agreements* refer to all consensus decisions made by member states, legally binding or not. Generally, an agreement is less formal than a treaty and addresses a more limited range of issues. *Resolutions* are more formal decisions used by UN organs, like the General Assembly, to express an agreement or a conclusion. *Programs or platforms (plans) of action* are blueprints for a series of actions that nations have agreed should be taken on a specific set of issues at national, regional and international levels. They are statements of political will and commitment, but are not legally binding. The plan of action that is the central subject of this chapter is the Madrid International Plan of Action on Aging (MIPAA), a plan passed by the General Assembly in 2002 to promote the well-being of older persons. *Declarations* lie between resolutions and treaties or conventions, and can carry the weight of customary law, such as the Universal Declaration of Human Rights. They "convey a high level of aspiration and political commitment, usually adopted at the Head of State of Government level" (Sidhu 2007, p. 46).

In contrast to agreements, resolutions, and plans of action, *treaties* refer to all instruments that are legally binding under international law. The contracting bodies must intend to create legal rights and duties; the instruments must be approved by States or international organizations with treaty making power; the treaty must be governed by international law; and it must be in writing. *Conventions* are synonymous with treaties and can cover all legally binding international agreements. An example is the Convention for the Elimination of All Forms of Discrimination Against Women (CEDAW).

An Overview of Aging Policies of the United Nations

Globally, in 2009 an estimated 737 million persons were aged 60 years or over. Their number is expected to increase to two billion by 2050. In 2009 the oldest old, those 80 years or over, made up 14% of those 60 years and older. The number of centenarians is growing even faster than the oldest old and is expected to increase nine fold by 2050, from approximately 454,000 in 2009 to 4.1 million in 2050 (United Nations 2009).

Life expectancy has increased dramatically all over the world; better health care, advances in medicine and technology and greater educational attainment have allowed more people in developed and developing countries to live 15–20 years past what has been traditionally considered retirement age. "Life expectancy at birth has gained more than 20 years since 1950, to reach its current level of 68 years. At current mortality rates, men surviving to age 60 can expect to live another 18 years and women surviving to age 60 can expect an additional 21 years to life" (United Nations 2009). Given the accelerated pace of aging in all countries, researchers have tended to focus on population aging rather than on the situation of older persons, and thus have not kept pace with the economic, social, and cultural situations of older persons adequately (United Nations General Assembly 2010).

The present section examines the history of aging policies at the United Nations. It briefly highlights and analyzes international agreements and treaties created to guide UN Member states in their implementation efforts to empower older persons and promote their rights. The basis for these policies varies from a social welfare to a human rights perspective, where the issues of older persons are mainstreamed in the development process. The following discussion underscores the significance of each agreement and the impact it has had on the road to the Madrid International Plan of Action on Ageing.

In 1948, shortly after the inception of the United Nations (UN), the Permanent Mission of Argentina was the first to raise awareness of the rights of older people; Argentina submitted a draft declaration on the rights of older persons to the General Assembly (Zelenev 2007). The declaration called for the "rights of older persons to assistance, housing, food, clothing, health care, recreation and work as well as stability and respect" (Zelenev 2007, p. 2). As a result of Argentina's draft declaration, the General Assembly adopted Resolution 213 on the rights of older persons.

From that moment on, the issue of older persons remained on the agenda of the United Nations. In 1969, the government of Malta reintroduced a declaration on the rights of older persons to the General Assembly (Zelenev 2007). This provided the momentum that was needed to reenergize the UN to further discuss and examine the dynamics of population aging in socio-economic development. During the 1970s, the General Assembly adopted resolution 3,137 entitled, "Questions of the Elderly and the Aged" and the World Population Conference adopted the "World Population Plan of Action". The document called for all UN member states to consider the implications of changing demographics of older persons in their development policies (United Nations Association in Canada 1999).

Several years of discussion and research by many UN member states and UN agencies resulted in the UN General Assembly calling for a world assembly on the elderly in 1982 in Vienna, Austria. During the "First World Assembly on Ageing" world leaders gathered to discuss the policy and development implications of global aging and draft the first international document on aging. That same year, the General Assembly endorsed the Vienna International Plan of Action on Ageing. The Vienna Plan became the first document to deal specifically with older persons, but most important, it attempted to serve as a guide to UN Member states for formulation and implementation of policies and programs on aging. It also sought to strengthen the policy-making capacities of member states (Sánchez Rivera 2010) and to help member states view aging populations in the context of development (United Nations 2000).

The Vienna Plan identified three priority areas: (a) the sustainability of development in a world where the population is increasing in age; (b) the maintenance of good health and well-being to an advanced age; and (c) the establishment of an appropriate and supportive environment for all age groups (Zelenev 2007). Significantly, the Vienna Plan mentioned the impact on socio-economic development and of negative attitudes and narratives against older persons: "Negative narratives of judging individuals in terms of their number of years rather than in terms of their capabilities when deciding when they should retire leads gradually to the process of isolation from society" (Sánchez Rivera 2010, p. 252).

Several nongovernmental organizations felt that the Vienna Plan fell short of fully including older persons in socio-economic development gains and policies. These nongovernmental organizations gathered to draft a new international document that would include older persons in socio-economic development and underscore the contributions older persons make to society. In 1991, the General Assembly adopted the United Nations Principles for Older Persons. The new programme included 18 principles that were grouped under five quality of life characteristics: independence, participation, care, self-fulfillment and dignity (Sánchez Rivera 2010).

Augmenting these UN initiatives, the global momentum was shifting away from a social welfare perspective, where government was seen as the purveyor of all needs, to a human rights perspective, where older persons were part of the development process. Within that spirit, in 1999 the General Assembly designated that year as the International Year of Older Persons. The General Assembly adopted the slogan "Towards a Society for All Ages" from the 1995 Summit for Social Development in Copenhagen, Denmark, for the International Year of Older Persons. This slogan signified that the issues of aging should be taken from a life course approach, and the benefits of socio-economic development should enhance the lives of all people, regardless of age. The General Assembly resolution noted that the concept of a Society for All Ages has four interlocking dimensions: the situation of older persons; lifelong development; multigenerational relationships; and the relationship between the aging of population and development (Zelenev 2007).

The message of a Society for All Ages and the International Year of Older Persons reverberated worldwide. The message affirmed that advancing the rights of old persons does not have to come at the expense of the younger generation (Sánchez Rivera 2010). Rather, the message is that the fruits of development and the inclusion of older persons, as stewards and benefactors of development, benefit all people throughout the life course.

The International Year of Older Persons and its political message paved the way for the preparation of the Second World Assembly on Ageing and the formulation of the Madrid International Plan of Action on Ageing. The purpose of the MIPAA was to reflect the new demographic realities of population aging and the inclusion of active aging.

The preparation for the Second World Assembly served to bridge many gaps in the understanding of a variety of issues related to population aging. The Second World Assembly assembled representatives of nongovernmental organizations, academics and policymakers on the new priorities facing older persons in the twenty-first century (Zelenev 2007). The areas of focus and discussion for the preparation became lifelong individual development; health and well-being; intergenerational equity; social protection; and enabling and supportive environments. Unlike the Vienna Plan, active aging was the guiding concept among the MIPPA priorities (Zelenev 2007).

The Second World Assembly on Ageing was held in Madrid, Spain in 2002 to mark the 20th anniversary of the First World Assembly on Ageing in Vienna. The purpose of the gathering was to review the implementation of the recommendations set forth by the First World Assembly on Ageing and the Vienna Plan. The goal of the groups and the committees involved in the formulation of the MIPAA was to create a new plan of action that reflected the new social, cultural, political, economic and demographic realities of the twenty-first century. The MIPAA sought to extend and enhance the initiatives of the Vienna Plan, such as significantly mainstreaming aging issues in all aspects of development (Sánchez Rivera 2010); interagency cooperation; and interdependence between generations (United Nations 2000).

The MIPAA has served as a guide to member states on their roles and responsibilities for upholding and enforcing the rights of older persons. It includes three priority areas for action: (1) older persons and development; (2) advancing health and well-being into old age, and (3) ensuring enabling and supportive environments.

This chapter will focus on Priority Areas 1 and 3 because of the relevancy of these two priority areas to the topic of age discrimination and creating enabling environments to promote positive images of older people, including older workers, and to address issues of abuse and mistreatment. These topics will be examined in the context of the Madrid International Plan of Action on Ageing and its implementation in social policies in the United States and Canada. Priority Area 2 of the Madrid Plan of Action focuses on older persons' health; while critical to the overall well-being of older persons, this priority area is not the focus of the book and will not be examined in this chapter.

An essential ingredient in the well-being of older workers the world-over is an enabling and supportive workplace environment in which the rights of older workers are respected. In keeping with MIPAA priority areas under discussion, the following provides an overview of some of the latest research and developments on ageism in the workplace in recent years, drawn from a broad cross-section of academic disciplines.

Older Persons and Development

As more countries develop socially and economically, it is easy for older persons to become marginalized, excluded and victimized by ageist attitudes in their communities. Advances in technology, concerns over social insurance and health care entitlements, and intergenerational fears over jobs are serious concerns that could have an impact on the social cohesion of society if not adequately addressed.

In response to these concerns the UN encourages members of civil society, states and the private sector to mainstream the needs of older persons and include them in the benefits of economic development. The UN defines mainstreaming as,

> ....a process of bringing issues that have not received the consideration they deserve to the center of attention. It is also a strategy and process as well as a multi-dimensional effort. Mainstreaming has been used as an advocacy tool for gender and human rights and the rights of the disabled. According to the UN successful mainstreaming should lead to (a) greater social integration of a particular group and (b) to the inclusion of a particular issue into all aspects of social, political, economic and cultural life. The overall objective of mainstreaming should be the achievement of more equitable development within a society for the benefit of *all* social groups (Venne 2005 p. 2).

For example, health, education, housing, transportation, and employment need to be viewed through the eyes of a particular social group, in this case older persons, to promote awareness of the specific concerns of that group, as well as more inclusive policymaking (Venne 2005). As our population continues to age, government should examine policies and programs that restrict older persons from fully benefiting from economic growth. Such policies include retirement rules that force older workers to leave the workforce upon qualifying for retirement benefits. They also include practices that deny older workers full access to education or training in formal or informal settings.

According to Kelly (2005),

> ...the World Health Organization affirms that countries can afford to get old if governments, international organizations and civil society enact active aging policies and programmes that enhance the health, participation and security of older citizens. These policies and programmes should be based on the rights, needs, preferences and capacities of older people. They also need to embrace a life course perspective that recognizes the important influence of earlier life experiences on the way that people age (p. 5).

Harnessing the strengths of older persons and their positive contributions to the workforce and society can not only secure the rights of older persons, but also provide economic benefits to the employer and the country.

Active Participation in Society and Development

"A society for all ages encompasses the goal of providing older persons with the opportunity to continue contributing to society. To work towards this goal, it is necessary to remove whatever excludes or discriminates against them" (United Nations 2003, p. 13).

Based on the theme of "Building a Society for All Ages", from the Second World Assembly on Ageing in 2002, many countries, including the United States and Canada, adopted the slogan as their policy development strategy on population aging. "A Society for All Ages" encompasses a responsibility by government and all of society for building a community where all people can fully participate and contribute their talents, skills, time and experience towards the betterment of society. Older persons contribute to society through paid employment. But their contributions are not only economic; they also include volunteer work, caring for friends or family, and mentoring.

The MIPAA and other international documents and treaties call for creating societies free of discrimination where all people can be contributing members of society. Nevertheless, the MIPAA is not a binding agreement and countries are not required to report on their implementation efforts. As a consequence, the implementation process varies with each country's current stage of economic and social development and extent of its aging population (United Nations General Assembly 2010).

Older persons in many parts of the world associate retirement with freedom, happiness and satisfaction. They want to remain active and positive in later years and continue to contribute to society. Many value family, friends and physical activity over money as a way to secure happiness in old age (Kelly 2005). The formulation and implementation of societal policies must integrate all groups, and remove obstructions—such as poverty, poor health, low educational attainment, limited access to transportation and age discrimination—which serve to exclude or marginalize older persons. Collectively, these policies may be called *social integration* policies.

Integrating older persons into public policy not only benefits society as a whole, but also provides economic and social benefits. For example, in the United States, older persons' contributions account for approximately $3 trillion a year in economic output (Brady and Coy 2005).

Social integration policies ensure that all people, regardless of age, ethnic background, educational attainment, disability or income level, have equal opportunities to contribute and participate in their communities. They are not only beneficial to the economy; they are also fundamental for social cohesion. Social cohesion affords individuals longer and healthier lives, and it affords communities greater economic contributions of its citizens. It also fosters contributions that cannot be measured in dollars and cents—such as caring for grandchildren or a loved one, and mentoring young people in school or at work.

Work and the Aging Labor Force

Priority Direction 1 of the MIPAA states that "Older Persons should be enabled to continue with income-generating work for as long as they want and for as long as they are able to do so productively" (United Nations 2003, p. 15). As the labor force ages in many parts of the world and in almost all industries, business leaders, human resource managers, government officials and policy makers are increasingly concerned about the shrinking labor force, and rising costs for health care, insurance, pensions and social security.

In the United States, the U.S. Congressional Budget Office predicted that labor force growth will slow by almost half over the next 10 years (Brady and Coy 2005). Certain skills and traits that older workers possess would be lost unless changes are made by employers and policymakers, to attend to these concerns, employers have begun to provide age discrimination training for all employees (Dennis and Thomas 2007). Since 2000, the Social Security Administration has allowed older workers to receive their full benefits regardless of how much they earn from working after age 65. The IRS has proposed similar rules where older workers could collect part of their pension if they continue to work past the age of 59½ (Johnson 2007).

Despite concerns from government and the private sector, many older persons want to work past retirement age, but they are often held back by ageist attitudes and policies in the workplace. These attitudes and policies can be subtle and pervasive. Negative stereotyping of older persons as rigid, incapable of learning, or lacking creativity has had an impact on how they are perceived by managers and colleagues (Dennis and Thomas 2007). To make matters worse, age discrimination laws are not always enforced and, in some cases, government agencies responsible for monitoring discrimination do not take cases against ageism seriously (Dennis and Thomas 2007).

International research has indicated that older persons do not feel old; although many associate old age with declining health, they do not believe that turning 60 or 65 makes one "old" (Bazalgette et al. 2011). Given the extensive experience and knowledge of older workers, it would be foolish for businesses to force older workers to leave. In Canada 80 years ago, life expectancy was 59 years old for men and 61 for women. Most workers died shortly after retiring. Today's retirees, on the other hand, can expect to live up to two decades after retirement (Duchesne 2004).

Recent years have seen an increase in older workers staying in the labor force. Many like to continue working because they love their jobs and feel they have much to contribute. Others work past retirement because they need the income (Duchesne 2004). The labor force participation rate for men age 55–64 has seen a continuous increase from 58% in 1995, to 67% in 2007 and in women, age 55–64 labor force participation rates increased from 35% in 1995 to 53% in 2007 (Ferrao and Marshall 2007).

In a survey by AARP, older workers were perceived as "experienced, knowledgeable, display positive work habits and commitment to quality, punctuality, ability to keep cool in crisis, respect for authority, solid performance, and basic skills in getting along with coworkers" (Dennis and Thomas 2007, p. 86). With these positive traits why do businesses hesitate to hire, promote and train older workers? Sometimes negative perceptions such as "inflexible, unwilling to adapt to new technology, lacking an aggressive spirit, resistant to new ways, and having some physical limitations" (Dennis and Thomas 2007, p. 86) can outweigh the abilities, energies and strengths of older persons for managers who make decisions about hiring, retaining, and promoting older workers.

Integrating older persons into the workplace and delaying retirement have many economic and social benefits. For example, a report by the Urban Institute concluded that delaying retirement can boost living standards for both workers and non-workers, generate additional tax revenue to fund government programs, increase retirement income, imbue the lives of older persons with meaning, and boost physical and emotional health by keeping older adults active and engaged (Johnson 2007). Expanding opportunities for older workers in the workforce can go a long way towards eliminating societal ageism because in modern societies paid labor, in large part, determines society's perception about the value of persons (Kelly 2005).

Despite pervasive age discrimination in the workplace, many employers acknowledge the experience and talents of older workers and most realize that replacing years of experience is difficult. Many companies have committed to keeping older workers in the workplace by implementing a number of changes. Some companies have used flexible working schedules, work during peak seasons, opportunities to work from home, and opportunities for continuing education.

In the United States, employers are also tapping the productivity of older workers. For instance, Con Edison in New York State is trying to retain its older workforce by implementing elder-care referral services and career-long training. In Massachusetts, MITRE Corp, a research and development company was concerned about losing workers with expertise in the radar field; to maintain its competitive edge, the company decided to bring back its retired workers on a part-time or on-call basis. Other companies, such as Deloitte Consulting LLC, have retained highly valued older employees by designating them as "senior leaders" and giving them incentives such as flexible hours and work locations, special projects, and opportunities to mentor younger workers and conduct research (Brady and Coy 2005).

Employment Opportunities for All Older Persons Who Want to Work

A recent AARP survey found that 68% of older workers (those 50 years of age and older) intended to work in retirement (Johnson 2007). In another survey of workers aged 51–56, Johnson (2007) reported that the proportion of workers interested in working past the age of 65 increased from 27% in 1992 to 33% in 2004. Many older workers want to work past the usual retirement age. Some want to continue working full time, while others want to work part-time or phase gradually into retirement. Unfortunately, according to AARP, these workers often face barriers such as age discrimination in the workplace, poor working conditions and stringent pension rules that penalize work past the age of 65.

Often employers feel that older workers no longer have the energy or capability to be productive employees. These ageist beliefs on the part of managers and employers may lead to terminating employees, demoting them to lower positions, denying them promotions and not inviting them to participate in training workshops. Such actions suggest that governments must take stronger action against age discrimination in the workplace.

Retirement schemes can discourage older persons from continuing to work past retirement age. In the United States, some employers provide employees with *defined benefit plans*. In a defined benefit plan, employers commit to paying their employees a specific benefit for life, beginning at retirement. Workers who participate in defined benefit plans are penalized if they choose to work past their retirement age. Some employers offer *defined contribution plans*. In these plans, benefit payments that are made to the employee are based on contributions made by workers during their time of employment, regardless of time of retirement. Because increased employee contributions yield increased post-retirement benefits, these plans encourage workers to work past retirement age. The US government has taken action to encourage older workers to stay in the labor force. For instance, the government has raised the age to collect Social Security benefits from 65 to 67 for workers born after 1959 (Johnson 2007).

Improving working conditions for older workers goes a long way toward increasing their productivity, energy and motivation. Developing better working conditions involves modifying the work space, and developing or implementing new tools, methods and equipment that meet the special needs and concerns of older workers (Kelly 2005). Sometimes modifications in the workplace may not be necessary at all; instead, changes in work schedules (such as part-time or flex-time options) are all that are needed to ensure older workers continue being productive members of society.

Access to Knowledge, Education and Training

Within the paradigm of lifelong learning, affirmed by international treaties and conventions, older persons can enjoy the right to an education, be it in formal or informal settings. Building the human resource capacity of older workers involves nurturing their potential, experience, skills and wisdom, and access to knowledge, education and training is a prominent part of the MIPAA in Priority area 1: older persons and development (UN 2003, p. 20). Opportunities for all workers to learn new technologies, employment methodologies, and job enhancing skills are invaluable in order to promote intergenerational solidarity and pride in being able to contribute to societal productivity.

If training in the workplace is fundamental to enhancing earnings and productivity, why are older persons often excluded by employers and mangers? A study of employers and their employees (Johnson 2007) shed light on this question. Johnson found that employees discerned training as an investment that would reap considerable earnings for the employer. Employers understood that their investment could be lost if employees leave. Thus, employers focused on profiting from employee training increasingly choose department-centered training, where loss would be minimized if an employee leaves.

Ageist attitudes of employers about the capability of older workers and imminent retirement often limit older workers' opportunities for trainings and continuing education (Johnson 2007). For example, Frazis et al. (1998) noted that 51% of workers ages 55 and older in large and medium sized establishments received formal employer-sponsored training during the previous 12 months, compared with 79% of those ages 25–34, 75% of those ages 35–44, and 65% of those ages 45–54, based on findings from a 1995 survey of employer-provided training conducted by the United States Department of Labor Bureau of Labor Statistics. Older workers also received less intensive training. According to the same survey, workers ages 55 and older averaged 17 h of informal on-the-job training between May and October, 2005, compared with 30 h for those ages 35–44 and 39 h for those ages 44–54 (Johnson 2007, p .25).

In 2011, The United Kingdom phased out the Default Retirement Age (DRA). The Default Retirement Age allowed employers to dismiss workers because they had reached the age of 65. An employer could fire an older worker based on capability but now, employer will need to objectively justify terminating employees 65 and older. In response to the dissatisfaction of employers, the government has developed a website for businesses to learn about ways they could retrain and trail all workers, in particularly older workers. The UK government recommends that all businesses should,

> ….ensure training for all employees regardless of their age; engage regularly with staff on an individual basis to discuss future training and promotion options; ensure all workers, including older workers, know what training opportunities are relevant to them and encourage them to take part; monitor outcomes from promotion exercises to ensure age bias has not occurred; and encourage mentoring and sharing of skills within your business-draw on experienced workers to support and train new staff members and other colleagues (United Kingdom Department for Work and Pensions 2011).

Intergenerational Solidarity

The 1995 UN Secretary-General's report defined the phrase "Society for All Ages" "as one that would encourage generations to invest in one another and share in the fruits of investment, guided by the twin principles of reciprocity and equity" (Kelly 2005, p. 33). The issue of intergenerational solidarity is included in the MIPAA under Priority area 1: Older people and development as fundamental to the realization of a society for all ages and to sustain economic growth and development that benefits all members of society (UN 2003, p. 22). However, some commentators believe that older generations are at odds with younger generations over finite government resources and policy attention. This conflict is often escalated by negative narratives in the media, portraying older persons as "selfish or greedy for government resources or employment opportunities" (Bazalgette et al. 2011, p. 37).

However, recent national surveys fail to confirm the contention that older generations are at odds with younger generations over public services. In the United States, many businesses acknowledge the contributions of older workers, their knowledge, skills and years of experience. "A relatively new awareness of older workers has emerged as retiring boomers take their knowledge and skills with them as they exit the front door. For example, companies such as Monsanto, the Aerospace Corporation, Cigna Insurance and CVS Pharmacy have established formal programs to rehire their retirees" (Dennis and Thomas 2007, p. 88). The retirees serve as mentors to younger staff members. Other businesses have modified their retirement policies by extending the retirement age and offering bonuses for delaying retirement. Such policies contribute greatly to the skill development of younger staff members and allow older persons to share their knowledge with others (Dennis and Thomas 2007).

Older persons have much to contribute as purveyors of knowledge and repositories of history, values and experience. The experiences and richness of the younger and older generations should not be broken by negative narratives by the media or policymakers; rather, each generation should nourish the other by supporting and working together for positive social change.

Priority Direction 3: Ensuring Enabling and Supportive Environments

In this priority direction, issues of elder abuse and mistreatment are presented in Issue 3: Neglect, abuse and violence (UN 2003, pp. 43–4). The introduction to this issue states that abuse against older people takes many forms, including physical, psychological, emotional, and financial, and gender affects risk in that older women are noted as experiencing a greater risk due to discriminatory societal attitudes and lack of human rights. It also notes that women's poverty can create vulnerability to abuse; and acknowledges that this is the result of economic opportunity, and lack of access to resources, education and support service. However, this section does not address barriers to productive employment based on discriminatory practices and abuse in the workplace faced by older adults.

Issue 4, Images of ageing (UN 2003, pp. 44–45) states that promoting positive images of older persons and ending ageist practices in the workplace is necessary to achieve social inclusion, lifelong development and independence. The best approach is to increase positive views of aging among the general public. For example, in recent years, the Canadian government has taken proactive steps to change the way older persons are perceived and treated by Canadian society: "The federal government leads an aggressive campaign to portray healthy aging and to present the benefits of staying active at all ages—in volunteer work, continuous learning and physical activity" (The Special Senate Committee on Aging 2009, p. 16). This campaign is targeted to schools from elementary to university level, media and the private sector to change the way older persons are perceived and treated by society. Other countries are well advised to follow Canada's lead in portraying positive images of older persons and their ability to contribute to society.

Conclusion

As the world population ages, employers, managers, policy makers, government officials and older persons themselves are challenged to re-think the role of older persons in the workplace. It is in the interest of all generations to avoid negative characterizations of older persons, and to view them as vital and productive members of society. Governments and international bodies, such as the UN, have played a role in promoting policies that support the productive efforts and opportunities of the older generation. But much is left to be done—by individuals, businesses, organizations and government—to ensure that older persons achieve their appropriate place in the world of work.

References

Bazalgette, L., Holden, J., Hubble, N., Morrison, J., & Tew, P. (2011). *Coming of age*. DEMOS. Retrieved on April 23, 2011, from http://www.demos.co.uk/files/Coming_of_Age_-_web.pdf?1302099024

Brady, D. & Coy, P. (2005). Old. Smart. Productive. Surprise! The greying of the workforce is better than you think. *Business Week*. Cover story, 1–5. Retrieved on April 23, 2011, from http://www.businessweek.com/print/magazine/content/05_26/b3939001_mz001.htm?chan=gl

Dennis, H., & Thomas, K. (2007). Ageism in the workplace. *The American Society on Aging, Generations* XXXI(1):84–89. Retrieved fromhttp://www.asaging.org/publications/dbase/GEN/Gen.31_1.Dennis.pdf

Duchesne, D. (2004). *More seniors at work*. Perspectives, Statistics Canada-Catalogue no.75-001-XIE, 5–16. Retrieved on May 5, 2011, from http://www.statcan.gc.ca/pub/75-001-x/10204/6789-eng.htm

Ferrao, V. & Marshall, K. (2007). *Perspectives on labour and income: Participation of older workers*. Statistics Canada. Ontario: Canada Government Printing Office. Retrieved on May 5, 2011, from http://www.statcan.gc.ca/pub/75-001-x/2007108/article/10303-eng.htm

Frazis, H., Gittleman, M., Horrigan, M., & Joyce, M. (1998). Results from the 1995 survey of employer-provided training. *Monthly Labor Review, 121*(6), 3–13.

Gray Panthers Action Network. (2012). Gray Panthers age and youth in action. Retrieved on March 12, 2012, from http://graypanthers.e-actionmax.com.

Johnson, R. W. (2007). *Managerial attitudes toward older workers: A view of the evidence*. The Urban Institute, The Retirement Project (Discussion Paper Series 07-05). Retrieved May 16, on http://www.urban.org/UploadedPDF/411548_managerial_attitudes.pdf

Kelly, P. (2005). *Integration and participation of older persons in development*. UN Department of Economic and Social Affairs, UN Secretariat (pp. 1–39). Retrieved on April 23, 2011, from http://www.un.org/ageing/documents/papers/Integration_participation.pdf

Office of the United Nations High Commissioner for Human Rights. (2003, November 10–18). *Substantive issues arising in the implementation of the international covenant on economic, social, and cultural rights*. Committee on Economic, Social and Cultural Rights. Thirty-Fifth Session, Geneva. Item 3 of the Provisional Agenda. Retrieved July 15, 2011, from http://www.unhchr.ch/tbs/doc.nsf/(Symbol)/b0756d872169c1a6c1256dd7003893f0?Opendocument

Sánchez Rivera, S. J. (2010). *Worldwide ageing: findings, norms, and aspirations*. Retrieved on June 14, 2011, from http://www.scribd.com/doc/39478479/11/C-An-Ageing-World-Population-and-Migration

Sidhu, G. (2007). *Intergovernmental negotiations and decision making at the United Nations: A guide* (2nd ed.). New York: United Nations Non-Governmental Liaison Service (NGLS).

The Special Senate Committee on Aging. (2009) *Canada's aging population: Seizing the opportunity*. Retrieved on May 5, 2011, from www.parl.gc.ca/40/2/parlbus/.../senate/.../AgingFinalReport-e.pdf

United Kingdom Department for Work and Pensions. (2011). *Age positive good practice case studies: Managing without a fixed retirement age*. London, UK. Retrieved on May 15, 2011, from www.dwp.gov.uk/.../good-practice-managing-without-fixed-retirement-age.pdf

United Nations. (2003). *Second world assembly on ageing, political declaration and Madrid international plan of action on ageing, 2002*. New York: United Nations Department of Public Information.

United Nations Association in Canada. (1999). *The United Nations and the question of ageing*. Retrieved on June 14, 2011, from http://www.unac.org/iyop/unquest.html

United Nations, Department of Economic and Social Affairs, Population Division. (2009). *Population ageing and development 2009*. Retrieved on June 14, 2011, from http://www.un.org/esa/population/publications/ageing/ageing2009chart.pdf

United Nations, Department of Economic and Social Affairs, Programme on Ageing. (2000). *Envisioning the 2nd world assembly on ageing: April 2002, Address by H.E. Julia, T. Alvarez, Ambassador/Permanent representative of the dominican republic to the United Nations, New York, From Vienna to Madrid*. Retrieved on July 15, 2011, from http://social.un.org/index/Ageing/InternationalDayofOlderPersons/2000/JuliaTAlvarezAmbassadorDominicanRepublic.aspx

United Nations General Assembly. (2010). *Follow-up to the second world assembly on ageing, report of the Secretary-General*. Sixty Fifth Session, Social Development: Follow-up to the International Year of Older Persons: Second World Assembly on Ageing, July 22, 2010. Retrieved on July 18, 2011, from www.un.org/ageing/.../A%2065_158/A%2065%20158%20En.pdf

Van Dooren, M. I. (2003). *The right to work, office of the high commission for human rights* (pp. 1–7). Retrieved on May 15, 2011, from http://www.unhchr.ch/tbs/doc.nsf/(Symbol)/b0756d872169e1a6c1256dd7003893f0?

Venne, R. (2005). *Mainstreaming the concerns of older persons into the social development agenda*. UN Department of Economic and Social Affairs, UN Secretariat (pp. 1–5). Retrieved on May 16, 2011, from http://www.un.org/ageing/documents/mainstreaming/positionpaper.pdf

Zelenev, S. (2007) The Madrid plan: A comprehensive agenda for an ageing world. UN Department of Economic and Social Affairs, Programme on Ageing, UN Secretariat. Retrieved on June 14, 2011, from social.un.org/index/LinkClick.aspx?fileticket=jpOj8HtfwHU%3D&tabid

Chapter 2
Definitions and Theoretical Models for Understanding Ageism and Abuse in the Workplace

Patricia Brownell and Mebane Powell

Introduction

- Sally, age 60, is told by her manager that her division is being reorganized and her job is being eliminated. When Sally asks if she can apply for a new job in the reorganized division, she is told the agency is looking for a Millennium type of worker. After leaving her job, Sally seeks treatment for depression and anxiety.
- Alex, age 67, is assigned to a project team with younger team members. The project requires expert computer skills. Alex asks permission to attend a computer training course, but his request is denied. The younger team members complain about Alex, who is called into his supervisor's office and given a warning that his skills are not keeping pace with the requirements of the job, that he is not getting along with his work colleagues, and that he may be terminated. Alex's job provides health benefits for his disabled wife.
- Wanda, age 73, overhears workers she supervises making fun of her appearance and the way she dresses, referring to her as an old bag and a fossil. She complains to her manager about their behavior, and is told that she is too sensitive, but that if she does not feel comfortable in the company she should think about taking a retirement package and leaving. Wanda lives alone and her job provides meaning in her life and needed financial support.

Gerontologists, policy makers, administrators, practitioners and researchers need to understand ageism, elder abuse and discrimination in the workplace. These features of the working lives of older adults, those 60 years of age and older, co-occur and

P. Brownell, Ph.D. • M. Powell, M.S.W. (✉)
Graduate School Social Service, Fordham University,
New York, NY, USA
e-mail: brownell@fordham.edu; mpowell@fordham.edu

P. Brownell and J.J. Kelly (eds.), *Ageism and Mistreatment of Older Workers: Current Reality, Future Solutions*, DOI 10.1007/978-94-007-5521-5_2,

interact to create challenges for those adults who choose to remain in the workforce late in life. An understanding of how these phenomena interact in the workplace is important in order to develop and implement effective policies, programs and organizational change that address these issues simultaneously.

There is currently a movement underway to recognize social problems as human rights issues. The Universal Declaration of Human Rights, adopted by the United Nations General Assembly in 1948, has become the guiding document for examination of the issue of ageism and abuse in the workplace (Wronka 2007). Article 23 of the Declaration states, "Everyone has the right to work, to free choice of employment, to just and favorable conditions of work and to protection against unemployment" (United Nations 2007, p. 8). Ageist practices and abuse in the workplace, also cited in Article 23, violate the notion of just and favorable conditions of work.

Madrid International Plan of Action on Ageing 2002

In April 2002, delegates of more than 160 governments, intergovernmental institutions and Non-Governmental Organizations (NGOs) came together at the United Nations Second World Assembly on Ageing in Madrid, Spain, to revise and update the 1982 Vienna Plan on Ageing. The Assembly's outcome document, the Madrid International Plan of Action on Ageing, commits governments to integrating the rights and needs of older people into national as well as international economic and social development policies (Global Action on Aging 2007; United Nations 2008).

One of the central themes running through the Madrid International Plan of Action on Ageing or MIPAA (United Nations 2003) is that of "ensuring the full enjoyment of economic, social and cultural rights and civil and political rights of persons and the elimination of all forms of violence and discrimination against older persons" (p. 12). This is accomplished by promoting a favorable attitude among employers regarding the productive capacity of older workers as being conducive to their continued employment and promoting awareness, including self-awareness, of their worth in the workplace (United Nations 2003).

The MIPAA established an agenda with a two fold purpose: (1) to promote the participation of older people in the economic development of nations and, (2) to ensure that older people share in the benefits of economic development. Priority direction I (Older persons and development) affirmed the United Nations Principles on the Rights of Older Persons, including the right to a life of dignity free of abuse. Priority direction II, Advancing health and wellbeing into old age, while important, is not a primary focus of this chapter. Priority direction III (Ensuring enabling and supportive environments) emphasizes that a positive view of ageing is an integral part of the Madrid International Plan of Action 2002.

In Priority direction III, one issue (Issue 3: neglect, abuse, and violence) focuses on elder mistreatment that can take many forms, including physical, psychological, emotional, and financial. MIPAA states that the impact of the trauma felt by older people in experiencing one or more of these forms of abuse can lead to shame and fear that may cause reluctance to seek help (UN 2003, p. 43). A second issue (Issue 4) focuses on images of aging. It declares that an important action step in ensuring enabling and supportive environments for older people is to "develop and widely promote a policy framework in which there is an individual and collective responsibility to recognize the past and present contributions of older persons, seeking to counteract preconceived biases and myths and consequently, to treat older persons with respect and gratitude, dignity and sensitivity" (UN 2003, p. 45).

In this chapter, the authors use a human rights framework to discuss ageism and abuse in the workplace. They review the current state of the literature on ageism and elder abuse, and apply this knowledge to manifestations of ageism and abuse in the workplace. Finally, they propose an expanded conceptual framework to include mistreatment and bullying of older workers, identify additional risk factors that can create vulnerability for older adult workers, and suggest directions for future research.

Socio-Demographic Trends in the United States

To understand the current magnitude of ageism and elder abuse in the workplace, it is first important to understand current and projected demographics. As scholars, professionals, advocates, and policymakers are well aware, by the year 2030 adults aged 65 and older will represent 19.3% or one-fifth of the population, a significant increase from the 12.6% of the population this age group represented in 2007 (Administration on Aging 2008).

Similar increases are also occurring in the workplace. The labor force participation of men aged 65 years and older increased from 17.9% in 2002, to 20.5%, in 2007, and 21.9% in 2009. The labor force participation of older women increased from 9.4% in 2000, to 12.6% in 2007, and 13.6% in 2009 (Administration on Aging 2010; U.S. Department of Labor, Bureau of Labor Statistics 2008a). Together, older men and women accounted for 4.2% of the 2009 workforce and by 2016 this number is projected to increase to 6.1% of the total workforce (U.S. Department of Labor, Bureau of Labor Statistics 2008b). The increase in labor force participation of individuals 65 and older can be attributed to changes in economic conditions and retirement policies.

At the same time, the United States Bureau of Labor Statistics projects a 6.9% decrease in the number of individuals aged 16–24 entering, or on the verge of entering, the workforce from 2006 to 2016 (U.S. Department of Labor, Bureau of Labor Statistics 2008b). Therefore, employers will need to retain older workers. These trends set the stage for intergenerational conflict and discrimination against older workers who, by choice or economic need, remain in the workforce past the age of 60. Intergenerational tensions may lead to age-based biases resulting in harassment, bullying, and demeaning of older workers.

Ageism

Ageism is a term coined by Dr. Robert Butler in 1968 to describe the "systematic stereotyping of and discrimination against older people because they are old, just as racism and sexism accomplished this with skin color and gender" (Butler 1989, p. 139). Gerontologists have divided ageism into four types. According to the International Longevity Center (2006),

> *Personal ageism* is defined as ideas, attitudes, beliefs, and practices on the part of individuals that are biased against persons or groups based on their older age. Examples include exclusion or ignoring older persons based on stereotypic assumptions, and stereotypes about older persons and old age. *Institutional ageism* is defined as missions, rules and practices that discriminate against individuals and/or groups because of their older age. Examples include mandatory retirement, and devaluing of older persons in cost-benefit analysis. *Intentional ageism* is defined as ideas, attitudes, rules, or practices that are carried out with the knowledge that they are biased against persons or groups based on their older age. Intentional ageism includes carrying out practices that take advantage of the vulnerabilities of older persons. Examples include marketing and media that use stereotypes of older workers; targeting older workers in financial scams; and denial of job training based on age. *Unintentional ageism* is defined as ideas, attitudes, rules, or practices that are carried out without the perpetrator's awareness that they are biased against persons or groups based on their older age. This type of ageism is also known as *inadvertent ageism*. Examples include absence of procedures to assist old and vulnerable persons, lack of built-environment considerations (ramps, elevators, handrails), and language used in the media (p. 21).

Other definitions of ageism include *positive ageism*. For example, Palmore (1990) defined ageism as "any prejudice or discrimination against or in favor of an age group" (p. 4). As examples of positive ageism, older persons might be seen as wiser, more capable of grasping the big picture as opposed to details of a specific problem or situation, and happier than younger persons. Assumptions based on positive or negative ageism may lead to responses that are biased (Palmore 1990).

Ageism and Elder Abuse

Ageism has not often been discussed as an underlying cause of elder abuse. However, personal and institutional ageism has been associated with elder abuse (International Longevity Center 2006). According to the late Robert Butler, ageism – also identified as age discrimination and age prejudice – is manifested in the home, institutions and the workplace, and is both a civil and human rights issue (International Longevity Center 2006). The Ontario Human Rights Commission noted that abuse of older adults occurs in large part due to negative attitudes towards older people (Canadian Network for the Prevention of Elder Abuse 2011).

Ageist behavior has been identified as a sub-set of age-differentiated behavior. Of identified types of age-differentiated behaviors, elder abuse is considered the most egregious type of hurtful behavior directed at older adults (Pasupathi and Löckenhoff 2004). Elder abuse, also called maltreatment or mistreatment, has been defined as harmful behavior directed toward older adults by family members or trusted others such as formal or informal caregivers (McDonald 2008).

Although elder abuse is often associated with care dependency of the victim, this is far from reality. Active aging, a new paradigm of aging for older adults, is emerging in both developed and developing countries across the world, and is now recognized as a normative stage in the aging process (Mellor and Rehr 2005). Most older people live active productive lives well into old age. This new paradigm of aging is sometimes associated with the United States baby boomers cohort, which has been identified as a transformative generation. Better educated than previous generations, in better health, and with a history of social activism and sense of entitlement, it has redefined old age to be a time of continued productivity (Butler 2008). According to the United Nations Human Rights Council (2011), "active and dignified ageing for older people requires reframing society's concept of ageing to put more focus on the continued participation of older persons in social, economic, cultural and civic life as well as their continuous contributions to society" (p. 1). Unfortunately abuse of older people by family members and trusted others may occur at any stage regardless of their life situation or status.

Types of elder abuse defined in the literature include physical, psychological, emotional, and financial abuse or mistreatment, and intentional and unintentional neglect. These definitions parallel definitions of ageism stated above, including the notion of intentional and unintentional acts; but the definition of neglect implies dependency and need for caregiving (National Research Council 2003). While workforce accommodations for people with disabilities are mandated by law and regulation, care dependency in older adults is associated with domestic settings.

Theoretical Model for Elder Mistreatment

Elder abuse was identified as a social problem of note by researchers, practitioners and policy makers in the 1970s, although considerable controversy remains among researchers of elder abuse and mistreatment about definitions (Pillemer et al. 2011). Physical and emotional or psychological abuse, neglect, and financial exploitation by trusted others are generally recognized as categories of elder abuse. However, self-neglect and crimes including scams perpetrated by strangers targeting older adults because of their perceived vulnerability are not universally accepted as elder abuse.

Currently, an expanded conceptual framework of elder abuse has emerged to bring consensus to this definitional divide among researchers, practitioners and policymakers. This expanded framework includes three dimensions. Elder abuse with stranger crimes against the elderly is one dimension, and self-neglect is a second of three dimensions of the emerging broader definition of elder abuse. The third dimension is identified as elder mistreatment.

The dimension identified as elder mistreatment includes the more traditional understanding of abuse against the elderly as perpetrated by trusted others in the home and in dependent care institutions like nursing homes (Dr. Georgia Anetzberger, President, November 23, 2011, National Committee for the Prevention of Elder Abuse, Personal Communication). This chapter includes mistreatment of older adults in the workplace as a third component of elder mistreatment. It also recognizes the new paradigm of aging as presented in the Madrid International Plan of Action on Ageing 2002.

As noted above, many older people are productive vital members of society, challenging the assumptions underlying much of the elder abuse scholarship and practice to date that old age is synonymous with decline and frailty. However, underlying negative societal and individual assumptions about aging and older people, also referred to as ageism, can result in mistreatment of older adult workers including bullying, harassment and discrimination.

A key element of mistreatment of older adults is that it is perpetrated by trusted others, and this can include workplace colleagues of older workers such as managerial, supervisory and peer staff (Dr. Georgia Anetzberger,, November 23, 2011, Personal Communication). The authors of this chapter argue that the dimension of mistreatment of older people in the workplace should be included in a conceptual model of elder abuse in order to better reflect the reality of productive and active aging.

Efforts to develop a comprehensive theory of elder abuse or mistreatment in domestic settings generally posit a transactional model that includes a victim of mistreatment, and a responsible actor or perpetrator, both of whom are the focus of analytic attention (National Research Council 2003). Factors relevant to model-building are: the interactions between the characteristics of the victim and of the abuser; and contextual factors, which include location, social relationship, and the broader socio-cultural context defined by socio-demographic variables such as race, ethnicity, religion, region and economic status.

According to the Panel to Review Risk and Prevalence of Elder Abuse and Neglect, a panel of experts on elder abuse convened by the National Research Council and sponsored by the National Institute on Aging, National Institutes of Health, and the U. S. Department of Health and Human Services (National Research Council 2003), elder abuse and mistreatment can be understood by means of

> ...... a model of transactional process unfolding over time as the elder person, his or her trusted other, and other interested parties (stakeholders) {become} concerned with the older person and his or her well-being in the context of changes in the physical, psychological, and social circumstances of the several parties as the result of the elder person's aging process and life course (p. 62).

The panel also suggests that this model is embedded in an environmental socio-cultural context, which provides opportunities to include ageism as a social value attached to the aging process (National Research Council 2003). The socio-cultural context may include the home and family or an institutional setting, the latter implying a high level of care dependency on the part of the older adult victim. In this conceptual framework, which is a risk model, the sociocultural context of the older adult and the potential abuser mediates individual factors of both that interact with status inequality, relationship type, and power and exchange dynamics that can lead to elder mistreatment.

The panel identified individual level factors for both older adult victims and perpetrators: social status (wealth, race/ethnicity, religion and kinship status); physical health status (chronic diseases or acute illness); personality characteristics; mental health (dementia, depression, social competence, intelligence and other personality characteristics); and beliefs and attitudes about aging, kinship and caregiving obligations. Status inequality factors include differences in gender, age, race/ethnicity, and education (National Research Council 2003). Relationship types include non-co-residential, co-residential, marital partnership or other kin. Power and exchange dynamics include negotiation of caregiving scripts and dependency of the victim on the abuser and/or the abuser on the victim. Based on research on care dependency and elder abuse, these dynamics can pertain to both older adult victims and abusers (Pillemer and Suitor 1992).

Early studies on elder abuse found that abusers tended to be financially dependent on the victim (e.g., Wolf et al. 1982). Other studies examined the dependency of victims on abusers and did not find greater dependency on abusers by victims compared with non-victims (Bristowe and Collins 1989; Homer and Gilleard 1990; Phillips 1983; Pillemer 1985; Pillemer and Finkenhor 1989; Pillemer and Suitor 1992; Reis and Nahmiash 1998; Wolf and Pillemer 1989).

In an elaboration of the model for institutional care providers of care dependent elderly, the socio-cultural context can include region of country, and state regulatory and legal environment. Social embeddedness variables can include organizational characteristics, and individual-level factors can include individual characteristics such as age, gender, and race/ethnicity (National Research Council 2003).

According to the panel (National Research Council 2003) and others, existing theoretical models of elder abuse and mistreatment often imply a level of care dependency on the part of older adult victims. However, elder abuse, as an outcome of ageism, can occur regardless of care dependency on the part of older adult victims, and can occur in settings other than homes or institutions as well. Although there is an assumption in much of the professional literature that older adults are out of the workforce, statistics show that a significant proportion of adults 60 years of age and older, remain in the workforce well past 60 years of age (National Association of Social Workers 2008).

A Theoretical Model for Workplace Bullying

A theoretical framework for the study and management of bullying in the workplace was proposed by Einarsen et al. (2003). They noted that gerontologists first became interested in workplace bullying in the 1980s. This came about because of the interest of Swedish family therapist, Heinz Heymann, who transferred an interest in family conflict to conflict in the workplace. Thus, the theoretical models of workplace bullying and elder abuse and mistreatment have many variables in common.

Like the theoretical model proposed by the panel to review risk and prevalence of elder abuse and neglect (National Research Council 2003), the workplace bullying framework also hypothesizes that the phenomena of interest are embedded in a socio-cultural context. The theoretical model on workplace bullying, however, emphasizes organizational factors over individual factors, and is more of a descriptive than a predictive model. For example, organizational factors inhibiting aggressive behavior interact with individual, social and contextual antecedents of aggressive behavior. Policy enforcement and organizational action – including tolerance and intolerance, social support and retaliation and retribution – are affected by, and in turn, affect the organization and the individual.

Bullying behaviors – as exhibited by the perpetrator and as perceived by the victim – as well as immediate behavioral and emotional reactions, are influenced by individual characteristics of the victim. These include demographic factors and social circumstances, as well as personality and personal history.

The workplace bullying model differs from the elder mistreatment model in several ways. These include: (1) characteristics of the perpetrator(s) are not noted in the workplace bullying model; (2) emphasis is placed primarily on the perception of the victim rather than actual events; and (3) outcomes of the bullying are not individualized, but instead are linked to organizational action. Einarsen et al. (2003) view bullying as related to personal or environmental factors and presence or lack of organizational inhibitors of bullying behavior.

This model of workplace bullying may be strengthened by including the perpetrator's individual characteristics, and the outcome of the interaction among victim, perpetrator, organizational environment and socio-cultural context. The model on elder abuse can be strengthened by broadening the contexts in which abuse can occur (to include workplace settings) and including a stronger assumption that the older person who is victimized is not necessarily dependent on the abuser. The authors advocate a stronger emphasis on the inclusion of socio-cultural values like ageism – which cuts across home, institutional and workplace settings – as well as race/ethnicity and socio-economic status. Inclusion of these variables can guide interventions. It will also lead to more robust models that can be tested to determine what combination of variables best predicts elder abuse and workplace bullying. Armed with this knowledge, policy makers can help society to restore the human rights of older workers, as delineated in the MIPAA (United Nations 2003).

Additional Intervening Variables

Researchers have shown that ageism is "pervasive in media, healthcare, education and advertising" (e.g., Dennis and Thomas 2007), and often co-occurs with sexism (Barnett 2005; Hatch 2005), and technological change (Cutler 2005). As a counterbalance to this, in 1967 the U.S. Congress recognized and addressed ageism in the workplace by enacting the Age Discrimination in Employment Act (ADEA). Although the ADEA acknowledged ageism as a social problem that disproportionately affects employees 40 years of age or older, it did not address ageism as a human rights issue. Rather, the ADEA is an employment law that addresses ageism from an economic standpoint in that back pay and future loss may be awarded. While the ADEA has a provision that notes workplace harassment is illegal and can be pursued under the ADEA, there is precedence in the law for not awarding emotional and compensatory damages resulting from workplace harassment,[1] which is not the case with lawsuits filed under Title VII of the Civil Rights Act of 1964 (Dennis and Thomas 2007). Although there has been progress in recognizing that Americans are working beyond the age of 65 (the age limit of 70 for protection under the ADEA was lifted in 1987), there has been little movement to reframe age discrimination from an economic to a human rights issue in North America. (A more in-depth discussion of legal and other legislative protection of older workers is provided in Chap. 9).

Abuse in the Workplace

The impact of workplace abuse, or bullying, is costly to society, organizations and the individual victim. While organizations experience higher turnover, absenteeism, lower performance and productivity, and higher health care costs, people who experience harassment in the workplace often show "cognitive effects such as concentration problems, insecurity and lack of initiative" (Hoel et al. 2003, p. 130). From a societal viewpoint, people experiencing workplace harassment are less likely to stay in the workforce and may use their retirement savings earlier than anticipated, which could lead to an increased need for social welfare programs.

Workplace abuse is defined in various ways by diverse disciplines, including public health, substance abuse, nursing, law, and education (Keashly and Jagatic 2003). In 2007, a nationally representative study conducted by the Workplace Bullying Institute defined workplace abuse to include "any or all of the following types of repeated mistreatment: sabotage by others that prevented work from getting done, verbal abuse, threatening conduct, intimidation, [and] humiliation" and found that current workplace abuse for those aged 50–64 was lower than for any other age group (10% vs. 19% for workers aged 18–29). However, 30% of the oldest age group reported a lifetime incidence of bullying [that] was unacceptably higher than any other age group (26% for those aged 30–49; and 19% for those aged 18–29) (Namie 2008).

Other definitions of workplace abuse include Brodsky's 1976 (cited in Keashly and Jagatic 2003, p. 33) definition of "repeated and persistent attempts by one person to torment, wear down, frustrate or get a reaction from another. It is treatment

[1] (Rivera Rodriguez v. Frito Lay Snacks Caribbean, 265 F.3d 15, 24 (1st Cir. 2001)), the Law Firm Passman & Kaplan, PC Attorneys Washington, District of Columbia noted that "[a]lthough hostile work environment claims are recognized under the ADEA, the court asserted it is well established that compensatory damages for emotional distress and pain and suffering arising from a discriminatorily hostile or abusive environment based on age are not allowed" (No Compensatory Damages Under ADEA, n.d., para. 5).

which persistently provokes, pressures, frightens, intimidates or otherwise discomforts another person" (p. 2). From an organizational perspective, Robinson and Bennett in 1995 (cited in Keashly and Jagatic 2003, p. 33) used the definition of "voluntary behavior that violates significant organizational norms and, in so doing, threatens the well-bring of the organization or its members, or both" (p. 555). Instrumental bullying is used to denote purposive bullying. There is a parallel to this in elder abuse definitions, where psychological abuse may be used as a tactic to obtain money or material goods (Quinn 2000).

Keashly and Jagatic (2003), in their review of recent studies of workplace abuse, noted the broad and overlapping nature of definitions of abuse. Using a broad definition, and relying on insurance records, they studied the incidence of workplace abuse in the U.S. Based on a small sample of 600 full-time workers, they found that the incidence of workplace abuse was 25% (one in four individuals). They also noted a more recent 2000 statewide Michigan study that found 59% of individuals in the workplace reported some form of emotionally abusive behavior. These rates are alarmingly high.

Co-occurrence of Ageism and Abuse: Directions for the Future

What is known about the co-occurrence of ageism *and* abuse in the workplace? To shed light on this issue, in November 2009, Powell conducted a search using Fordham University's multi-database search option. The search terms "ageism and workplace abuse," "ageism and abuse," "ageism and bullying," "ageism and workplace bullying," "workplace abuse and older adults," "bullying and older adults" and "workplace harassment and older adults" were entered. Although many articles addressed the issue of abuse or ageism, few addressed the co-occurrence of workplace ageism and abuse in the Unites States. The findings suggest there is a gap in the literature that future researchers should address.

Researchers who study the co-occurrence of ageism and abuse in the workplace should strive for a definition of abuse that is clear and concise. In this way, they will help organizations to address the issue and to measure the impact and outcome of policy implementation and change. Such a concise, agreed-upon definition, achieved perhaps through consensus by a panel of experts, would also allow researchers to create a measure to capture the personal and organizational impact of abuse and ageism in the workplace. Most important, a definition of workplace ageism and abuse should draw on workers' own definitions. This will redress the lack of attention to older workers' perspectives in the workplace bullying conceptual framework discussed above (Powell 2010). This will also be a step toward ensuring implementation of the international ageing agenda in MIPAA 2002 so that older workers are no longer subjected to a denial of their basic human rights.

Conceptual models of abuse in the workplace (bullying), elder abuse, and domestic violence do exist and have several factors in common: power and control, dependency, and negative outcomes. It is time to revisit these models and to create a new conceptual model for ageism and abuse in the workplace. This would allow for a more comprehensive framework for investigating ageism and abuse in the workplace. Such a framework should address whether particular organizational structures promote or prevent workplace abuse and ageism.

Discussing the overlap of ageism and abuse will raise awareness and sensitize social workers, human resource administrations, and older workers themselves to the issue of workplace abuse. Through ongoing and open discussion, scholars, professionals, advocates, and students can arrive at a concise and agreed-upon definition for workplace abuse; a conceptual model that integrates ageism and abuse in the workplace; and research that explores the combined impacts of ageism and abuse. These steps will lead to effective workplace policies and practices that promote the human rights of older workers.

References

Administration on Aging. (2008). *A profile of older Americans: Employment*. Retrieved on December 15, 2011, from http://www.aoa.gov/AoARoot/Aging_Statistics/Profile/2008/12.aspx

Administration on Aging. (2010). *A profile of older Americans 2010*. Retrieved on December 15, 2011, from Http://www.aoa.gov/AoARoot/aging_Statistics/Profile/2011/

Barnett, R. (2005). Ageism and sexism in the workplace. *Generations, 29*(3), 25–30.

Bristowe, E., & Collins, J. B. (1989). Family-mediated abuse of non-institutionalized elder men and women living in British Columbia. *Journal of Elder Abuse & Neglect, 1*(1), 45–54.

Butler, R. N. (1989). Dispelling ageism: The cross-cutting intervention. *The Annals of the American Academy of Political and Social Science, 503*, 138–147.

Butler, R. N. (2008). *The longevity revolution: The benefits and challenges of living a long life*. New York: Public Affairs.

Canadian Network for the Prevention of Elder Abuse. (2011). *Ageism*. Retrieved from internet on October 2, 2011, from http://www.cnpea.ca/ageism.htm

Cutler, S. J. (2005). Ageism and technology. *Generations, 29*(3), 67–72.

Dennis, H., & Thomas, K. (2007). Ageism in the workplace. *Generations, 31*(1), 84–89.

Einarsen, S., Hoel, H., Zapf, D., & Cooper, C. L. (2003). The concept of bullying at work. In S. Einarsen, H. Hoel, D. Lapf, & C. L. Cooper (Eds.), *Bullying and emotional abuse in he workplace: International perspectives in research and practice*. London: Taylor & Francis.

Global Action on Aging. (2007). *The five year review of MIPAA*. Retrieved from the Internet on December 5, 2011, from http://globalaging.org/agingwatch/events/CSD/MIPAA+5.htm

Hatch, L. (2005). Gender and ageism. *Generations, 29*(3), 19–24.

Hoel, H., Sheehan, M., Cooper, C., & Einarsen, S. (2003). Organizational effects of workplace bullying. In S. Einarsen, H. Hoel, D. Zapf, & C. L. Cooper (Eds.), *Bullying and emotional abuse in the workplace: International perspectives in research and practice* (pp. 129–148). London: Taylor & Francis.

Homer, A. C., & Gilleard, C. (1990). Abuse of elderly people by their caregivers. *British Medical Journal, 301*(6765), 1359–1362.

International Longevity Center. (2006). *Ageism in America* (p. 21). New York: Open Society Institute.

Keashly, L., & Jagatic, K. (2003). By any other name: American perspectives on workplace bullying. In S. Einarsen, H. Hoel, D. Zapf, & C. L. Cooper (Eds.), *Bullying and emotional abuse in the workplace: International perspectives in research and practice* (pp. 31–91). London: Taylor & Francis.

McDonald, L. (2008). Editorial. *Indian Journal of Gerontology*. Special Issue. Jaipur: Aalekh Publishers (i–x).

Mellor, M. J., & Rehr, H. (2005). *Baby boomers: Can my eighties be like my fifties?* New York: Springer Publishing company.

Namie, G. (2008). *U.S. hostile workplace bullying survey 2000*. Bellingham: The Workplace Bullying and Training Institute.

National Association of Social Workers. (2008). *Aging and wellness* (pp. 14–21). Social Work Speaks. Washington, DC: NASW.

National Research Council. (2003). *Elder mistreatment: Abuse, neglect, and exploitation in an aging America*. Washington, DC: The National Academies Press.

Palmore, E. D. (1990). *Ageism, negative and positive*. New York: Springer.

Pasupathi, M., & Löckenhoff, C. E. (2004). Ageist behavior. In T. D. Nelson (Ed.), *Ageism: Stereotyping and prejudice against older persons* (pp. 201–246). London: The MIT Press.

Phillips, L. R. (1983). Abuse and neglect of the frail elderly at home: An exploration of theoretical relationships. *Journal of Advanced Nursing, 8*, 379–392.

Pillemer, K. A. (1985). The dangers of dependency: New findings on domestic violence against the elderly. *Social Problems, 33*(2), 146–158.

Pillemer, K. A., & Finkenhor, D. (1989). Causes of elder abuse: Caregiver stress versus problem relatives. *The American Journal of Orthopsychiatry, 59*, 179–187.

Pillemer, K. A., & Suitor, J. J. (1992). Violence and violent feelings: What causes them among family caregivers? *Journal of Gerontology, 47*, S165–S172.

Pillemer, K. A., Breckman, R., Sweeney, C. D., Brownell, P., Fulmer, T., Berman, J., Brown, E., Laureano, E., & Lachs, M. S. (2011). Practitioners' views on elder mistreatment research priorities: Recommendations from a research-to-practice consensus conference. *Journal of Elder Abuse & Neglect, 23*(2), 115–126.

Powell, M. (2010). Ageism and abuse in the workplace: A new frontier. *Journal of Gerontological Social Work, 53*(7), 654–665.

Quinn, M. J. (2000). Undoing undue influence. *Journal of Elder Abuse & Neglect, 12*(2), 9–17.

Reis, M., & Nahmiash, D. (1998). Validation of the indicators of abuse (IOA) screen. *The Gerontologist, 38*(4), 471–480.

United Nations. (2003). *Political declaration and Madrid international plan of action on ageing, second world assembly on ageing, Madrid, Spain, 2002*. New York: United Nations Department of Public Information.

United Nations. (2007). *Universal declaration of human rights: Dignity and justice for all of us, 60th anniversary special edition* (p. 8). New York: United Nations Department of Public Information.

United Nations. (2008). *Guide to the national implementation of the Madrid international plan of action on ageing*. New York: Department of Economic and Social Affairs.

United Nations Human Rights Council. (2011). *Thematic study on the realization of the right to health of older persons by the Special Rapporteur on the right of everyone in the enjoyment of the highest attainable standard of physical and mental health, Anand Grover*. United Nations General Assembly, Human Rights Council, Eighteenth session, A/HRC/18/37.

U.S. Department of Labor, Bureau of Labor Statistics. (2008a). *Older workers chart data*. Retrieved on December 15, 2011, from http://www.bls.gov/spotlight/2008/older_workers/data.htm#chart_02

U.S. Department of Labor, Bureau of Labor Statistics. (2008b). *Spotlight on statistics: older workers*. Retrieved on December 15, 2011, from http://www.bls.gov/spotlight/older_workers

Wolf, R. S., & Pillemer, K. A. (1989). *Helping elderly victims: The reality of elder abuse*. New York: Columbia University Press.

Wolf, R. S., Strugnell, C. P., & Godkin, M. A. (1982). *Preliminary findings from three model projects on elderly abuse*. Worcester: University of Massachusetts Medical Center, Center on Aging.

Wronka, J. (2007). *Human rights and social justice: Social action and service for the helping and health professions: Oakland*. Thousand Oaks: Sage.

Part II
Current Reality

Chapter 3
Harassment of Older Adults in the Workplace

Amy Blackstone

How do workers come to define certain experiences as workplace harassment and how do they respond to such experiences? How do age, gender, and power across domains such as work, family, and community life shape victimization experiences and mobilization, or actions, in response to harassment? Over time, scholars in diverse specializations such as work, aging, law, gender, and criminology have addressed these and similar questions. In this chapter, I combine these diverse perspectives in order to assess how age, gender, and other dimensions of power shape older workers' harassment experiences.

Older Workers in the United States

It is estimated that by the year 2012, approximately 20.8% of men and 12.1% of women aged 65 and over will be participating in the U.S. labor force, up from 16.3% of men and 8.6% of women in 1990 (U.S. Bureau of the Census 2006, p. 387). The projected increase in labor force participation among older workers is explained by the aging of the baby boomer generation. Not only are baby boomers getting closer to traditional retirement age, they also appear likely to continue their labor force participation longer than their predecessors did. Although labor force activity declines after age 62 (National Institute on Aging NIA 2007), a recent survey by the

An earlier version of this chapter was presented at the 2010 meeting of the American Sociological Association. Direct correspondence to amy.blackstone@umit.maine.edu. This research was supported by a grant from the National Science Foundation (SES-0817673). Many thanks to Mahala Stewart and Michael J. Costello for research assistance.

A. Blackstone, Ph.D. (✉)
Department of Sociology, University of Maine,
5728 Fernald Hall, Orono, Maine 04469
e-mail: Amy.blackstone@umit.maine.edu

P. Brownell and J.J. Kelly (eds.), *Ageism and Mistreatment of Older Workers: Current Reality, Future Solutions*, DOI 10.1007/978-94-007-5521-5_3,

American Association of Retired Persons (AARP 2004) found that 8 out of 10 baby boomers expected that they would continue working, at least part time, after they reached the age of initial eligibility for Social Security benefits.

As the nation confronts an aging workforce, so too does it face an increasingly broader class of workers for whom the United States Equal Employment Opportunity Commission (EEOC) provides the right to protections (Wakefield and Uggen 2004). In 2004 the EEOC reported having received 17,837 complaints filed under the Age Discrimination in Employment Act (U.S. EEOC 2006). The number of age-based complaints rivaled the number of charges filed under the Americans with Disabilities Act (15,376 in 2004) and was not far behind the more common categories of sex- and race-based complaints (24,249 and 27,696 in 2004, respectively). These statistics matter for any study of power in the workplace because, as two prominent scholars of aging recently put it, "Old people lose authority and autonomy" (Calasanti and Slevin 2006, p. 6).

Such estimates and projections have prompted public, political, and scholarly discussion of the unique health and safety issues facing aging workers and their employers (e.g., National Institute for Occupational Safety and Health 2004) and also the potential benefits of an aging workforce (e.g., Nacelewicz 2005; Sterns and Miklos 1995). Much research in the area focuses on employers' and coworkers' attitudes toward older workers and the impact of an aging workforce on the structure of the workplace (e.g., Brooke and Taylor 2005; Duncan 2003; Marshall 2001; Pitt-Catsouphes 2005; Rix 2001; Siegenthaler and Brenner 2000; Sterns et al. 1994; Taylor and Walker 1998); access to employment (e.g., Bendick et al. 1999; Hirsch et al. 2000); and workers' plans for retirement and reasons for later-life labor force participation (Appold 2004; Eichar et al. 1991; Ekerdt et al. 2000; Singh and Verma 2003). Other research studies focus on ageism but not on workplace sexual harassment or bullying per se (Dennis and Thomas 2007; Macnicol 2006; McMullin and Marshall 2001).

All these areas of inquiry are important. Yet so too is consideration of workers' own perceptions of their experiences. Although some scholars have studied older workers' perceptions of workplace discrimination (e.g., Altschuler 2004; Mor-Barak 1995; Tougas et al. 2004), these scholars have not considered the broader theoretical implications of their findings. In an AARP study of persons aged 55 and above, some participants wished to work but were not seeking employment because they feared they would be perceived as "too old" by employers, or face other forms of discrimination. The researchers concluded that "age continues to work against many older jobseekers" (Rix 2005, p. 3).

A number of questions remain unanswered. How does age impact workers' vulnerability to harassment and discrimination? What are workers' perceptions of their own vulnerability? And how do these perceptions change for workers as they gain experience in the labor force?

Insights from several areas of inquiry framed this study. Much knowledge has been gained about power in the workplace thanks to the efforts of feminist and gender scholars. Scholars of aging provide important additional insight into the experiences and perspectives of older individuals. Knowledge about workplace violence and victimization comes from scholars in the field of criminology, while understandings of legal consciousness and mobilization in response to harassment have been informed by socio-legal studies. Below I describe insights from these areas of inquiry with particular attention to how each area applies to the present study of harassment of older workers. In addition to the areas already mentioned, I also consider insights from life course studies and the sociology of work.

Power in the Workplace: Diverse Perspectives

Age, gender, social class, and access to social networks and support are among the most salient determinants of power shaping workplace interactions and experiences (see, e.g., Acker 1990; Calasanti and Slevin 2001, 2006; Kanter 1977; Rospenda et al. 1998; Uggen and Blackstone 2004). In the following section I draw from diverse perspectives to describe the dynamics of power relations and harassment in the workplace.

Within the scholarly community, gender researchers were among the first to draw attention to issues of power in the workplace. As the feminist movement of the 1960s and 1970s took shape, activists and scholars brought into public view what many knew privately, but did not discuss: some workplaces were fraught with problematic sexualized interactions shaped by a hegemonic gender ideology in which men asserted and maintained sexual and other dominance over women colleagues. These kinds of workplace interactions were labeled sexual harassment. Catherine MacKinnon's 1979 book, *Sexual Harassment of Working Women,* helped push the issue further into public view and placed it on the political and legal agenda. Sociocultural work in this area highlighted gender-based power differentials as the primary predictor of sexual harassment victimization (Welsh 1999). Although the research findings I present in the following sections of this paper are not limited to sexual harassment only, the insights of studies confined to sexual harassment should not be overlooked. This large and diverse body of literature offers general insight into power dynamics in the workplace, in part because the focus of much sexual harassment research has broadened over the years.

Sexual harassment researchers initially focused on gendered dimensions of power in the workplace, but the literature has since grown to include inquiry into the relationships between and among intersecting dimensions of power. For example, Rospenda et al. (1998) drew attention to the multiple levels at which power plays out in the workplace, arguing that the effects of gender, race, and class operate simultaneously, at cultural, organizational, and individual levels to shape individuals' workplace experiences. Additional research supports the idea that together gender, race, and class shape workplace harassment experiences (Giuffre and Williams 1994; Texeira 2002) as does sexual identity (DeSouza and Solberg 2004) and citizenship status (Welsh et al. 2006). Increasingly, sexual harassment researchers are including consideration of age as an additional dimension of power shaping workplace experiences.

The literature on age and work has a long history (e.g., Brooke and Taylor 2005; Marshall et al. 2001; Mortimer 2003) but within the context of workplace harassment, age has typically been left out of consideration. Studies of workplace sexual harassment that do analyze the impact of age on experiences or perceptions focus almost exclusively on mid-life or younger workers (Fineran 2002; Ford and Donis 1996; Padavic and Orcutt 1997; Uggen and Blackstone 2004) or rely on hypothetical scenarios rather than actual experiences (Wayne 2000). Other studies, focusing more generally on age as a dimension of power, examine its impact within various domains of social life, but not necessarily within the context of paid employment (e.g., Hess 1990; Levy 1988). Broader treatments of the impact of work on older adults have been published (e.g., Schaie and Schooler 1998) but have been criticized for not fully integrating the interaction of age with additional power dimensions such as race, class, and gender (Calasanti 1999).

A growing literature considers how age intersects with other power dimensions to influence work experiences and patterns (e.g., Barnett 2005; Calasanti 2002; Calasanti and Slevin 2001, 2006; McMullin and Marshall 2001; Smith and Calasanti 2005). Dominant conceptualizations of work itself highlight the connections between age and other dimensions of power. As Calasanti and Slevin (2001) noted, "productivity" should not be equated with paid labor alone, and older volunteers and others who participate in informal economies make this clear. Despite these advances in the conceptualization of age and power, Allen and Walker (2006), in an article on the importance of studying age, claimed that, "Much more integration is needed to connect a gender lens and an aging lens across disciplines…" (p. 155). McMullin (2000) suggested that age relations typically have not been considered in the social inequality literature alongside other dimensions of power such as gender, race, class, and sexual orientation. (See also King 2006). Finally, McMullin and Berger (2006) concluded that "relative to sexism, very little is known about the mechanisms through which ageism is manifested in paid work environments" (p. 219).

In addition to considering the literature on gender, it is also important to integrate the aging literature with insights from other disciplines, including criminology and law. Bringing criminological insights together with those from the study of gender, De Coster and her colleagues (1999) were among the first to test how well a general theory of victimization explains sexual harassment, applying routine activities theory to show that workplace victimization can be attributed to the presence of motivated offenders, the proximity of vulnerable targets, and the absence of solidaristic, or cohesive, work groups. Routine activities and other criminological theories focusing on motivated offenders and organizational culture point to the importance of analyzing the relationships between individual-level characteristics and the structural and institutional contexts within which individuals operate (Cohen and Felson 1979; De Coster et al. 1999; Knefel et al. 2004; Mueller et al. 2001). These theories are therefore useful for examining the role that age plays in shaping workers' experiences.

Researchers in this area also have described the dynamics and features of workplace violence and harassment. Building from their own and others' work on harassment and generalized workplace abuse (Fendrich et al. 2002; Richman et al. 1999; Rospenda 2002; Rospenda, et al. 2000), Rospenda and Richman (2004) developed a typology of generalized workplace harassment which includes covert hostility, verbal hostility, manipulation, and physical hostility. Others have studied "workplace abuse," expounding upon bullying and mobbing, two of the more dominant forms of abuse (e.g., Koonin and Green 2004).

In addition to these areas of inquiry, the present study also drew from the socio-legal perspective, which provides insight into individuals' perceptions of, and responses to, harassment. Theories of legal consciousness point to cultural and social-psychological factors in shaping perceptions of justice and responses to perceived injustices (Bumiller 1988; Ewick and Silbey 1998; Felstiner et al. 1980–81; Kritzer et al. 1991; Merry 1990). According to these theories, cultural schema, as well as social resources, shape individuals' understandings of law and legality.

Ewick and Silbey (1998) suggested that legal consciousness can take several different forms. One form causes individuals to perceive the law and legality as all-powerful and impenetrable, something over which they have little control. Others might believe that law and legality are a series of meaningless rules and therefore something to be manipulated. Finally, some view the law as a power to be avoided or resisted (see also Ewick and Silbey 2003). The concept of legal consciousness allows for the expression and examination of legality in ways not officially recognized by the law. Although it may be typical to think of the law as being enacted in formal, traditional ways (e.g., through lawsuits), law can also be invoked in informal, less official ways. A growing body of literature addresses various forms of legal consciousness and the contexts within which it is manifested (e.g., Connolly 2002; Fleury-Steiner 2003; Hoffman 2005; Hull 2003; Marshall 2005; Marshall and Barclay 2003; Richman 2001). However, the question of how or whether legal consciousness varies by age has not yet been examined.

Not only can legal consciousness take a variety of forms, but the invocation of the law and legality occurs in different ways (Hoffman 2003). Further, because research on aging populations shows that many are reluctant to acknowledge the impact or existence of ageism in their own lives or to self-identify as "old" (e.g., Fairhurst 2003; Minichiello et al. 2000), it is possible that older workers will *not* choose to identify their experiences either as harassment or as having anything to do with their age. Thus, mobilization in response to harassment, and as shaped by legal consciousness, may work differently for differently positioned workers. For example, workers with more resources to draw from, such as positive coworker, friend, or family relationships, may respond differently to harassment than those with fewer sources of social support. I consider these and other predictions in the form of "objectives questions" below.

Objectives Questions

This study is framed by four objectives, each presented in the form of a question. The study rests on the assumption that workers in the United States operate within a culture where particular dimensions of power, such as age and gender, are especially salient; youth is valued over age (Levy 1988); and hetero-normative masculinity is placed hierarchically above all other expressions of gender (Blackstone 2003). In the workplace especially, "old people face repudiation from younger people as being weak, sick, and unable to learn and as nearing death" (King 2006, p. 54). Within this context, workers hold variable amounts of power across several domains including the workplace, family life, and community life. In the present study I addressed the following questions about older workers' harassment experiences:

1) What is the content of older workers' harassment experiences?
2) Which older workers are most likely to become targets of workplace harassment?
3) How do older workers respond to potentially harassing behaviors?
4) What do older workers themselves have to say about their workplace experiences?

Data and Measures

The data come from a sample of workers identified with the assistance of one state-level division of a national program designed to place older workers by providing training and matching older workers with potential employers. This division serves approximately 380 clients each year and tracks and maintains contact information for past clients. Surveys were sent to 576 eligible past and current program participants who had been in the paid workforce at some point since turning age 62. Of these 576 individuals, 183 workers completed the survey, yielding a response rate of 32%. Although a variety of techniques were used to ensure a high response rate,[1] it is possible that one lesson to come from this research is that mailed questionnaires are not the most effective method of data collection for an older population, particularly when participants are asked to share details about workplace experiences that they may fear put them at risk of losing needed employment.

The age range of those who returned the survey was 62–87 years old; the mean age of participants was 70 years old and the median age was 68.5 years old. Seventy-six per cent of participants were women and 96% were white. The gender and race composition of participants reflected that of all program participants in the state where data were collected. Thirty-one per cent of participants had graduated from high school, 30% had completed some college, and 8% had earned a bachelor's degree.

[1] All potential participants were notified in advance of the survey mailing via an article describing the research and the planned mailing in the program's monthly newsletter. A $1 bill was included with each survey in the initial mailing to provide an incentive and an advance token of thanks to participants for returning the surveys. Two months after the initial mailing went out, those who were sent a survey were contacted by phone. Although returned surveys did not contain any identifying information about participants, research assistants contacted individuals to whom a survey had been mailed to remind them that it was not too late to return their survey and to say thanks to those who had already done so. Four months after the initial mailing, everyone on the original mailing list received a letter thanking those who had returned the survey and once again reminding those who had not, that it was not too late to do so. The letter included a return postcard for participants to complete should they wish to receive another copy of the survey. Participants were also provided a telephone number to call and the option of completing the survey by phone.

The primary focus of the survey was a set of workplace harassment questions, based on indicators used in other researchers' studies of generalized harassment (Koonin and Green 2004; Rospenda and Richman 2004) as well as indicators used in my own prior research on workplace harassment (McLaughlin et al. 2008; Uggen and Blackstone 2004). Additional questions, including who participants notified about harassment (e.g., no one, friends, family, coworkers, supervisors, attorneys, or government authorities) were placed on the survey, as was demographic and other background information. I also collected data on other aspects of participants' lives such as closeness in their relationships with family and friends. In addition to the quantitative survey questions, two open-ended questions were included on the survey in which participants were asked whether they believed any of their harassment experiences had anything to do with their age or gender (and to explain if so). A final open-ended question asked participants to describe what they had most enjoyed about their work experiences since turning 62.

Results

Objective 1: What Is the Content of Older Workers' Harassment Experiences?

Table 3.1 presents participants' reports of their harassment experiences.

As shown in Table 3.1, the four most common behaviors that participants said they encountered at work were having their work contributions ignored (25.1%); being left out of decisions that affect their work (23.0%); being talked down to by coworkers (20.8%); and being talked down to by bosses (20.2%). All four of these experiences indicated older workers' sense of being devalued in the workplace.

In my prior collaborative research on younger workers (Uggen and Blackstone 2004), my co-investigator and I asked about behaviors similar to those included in the present older worker survey. There were notable differences in the reports of younger and older workers. For example, younger workers most commonly experienced various kinds of offensive joking. Also, experiencing unwanted staring or invasion of personal space was much more common among younger workers. In both the younger and older worker samples, sexual assault and unwanted touching were rare, but both experiences were even less common among older workers. These patterns suggest that even though workers of all ages may be vulnerable to harassment, younger and older workers differ in the types of harassment they experience. The ways that workplace harassment manifests itself may change as workers age.

Additional behaviors that at least 10% of participants reported included verbal exchanges characterized by yelling or swearing; offensive age-related joking about the participant; offensive age-related joking about others; offensive sexual joking about others; being isolated from important work activities; comments or behaviors that demeaned participants' age; and unwanted questions about participants' private lives.

Table 3.1 Percent of older workers who experienced various types of harassing behaviors[a]

At any job you have held ***since turning 62****, have you experienced…*	**% Yes**
…Verbal exchanges characterized by yelling or swearing	**13.1%**
…Being talked down to by coworkers	**20.8**
… Being talked down to by customers	8.2
…Being talked down to by bosses	**20.2**
…Offensive age-related joking about you	**9.8**
…Offensive age-related joking about others	**11.5**
…Offensive sexual joking about you	2.7
…Offensive sexual joking about others	**10.4**
…Having your work contributions ignored	**25.1**
…Subtle or obvious threats to your safety	3.3
…Bribes to do things that are wrong	1.6
…Being hit, pushed or grabbed	2.7
…Being isolated from important work activities	**12.6**
…Being left out of decisions that affect your work	**23.0**
…Comments or behaviors that demean your gender	5.5
…Comments or behaviors that demean your age	**12.6**
…Unwanted questions about your private life	**10.4**
…Staring or invasion of your personal space	7.7
…Unwanted touching	1.6
…Sexually explicit pictures, posters, or other materials	1.1
…Sexual assault by a coworker or boss	1.1

[a]Experiences reported by at least 10% of participants are indicated in **boldface** type

Objective 2: Which Older Workers Are Most Likely to Become Targets of Workplace Harassment?

To understand which older workers were most likely to become targets of workplace harassment, I conducted *t*-tests of differences in responses to the harassment indicators between the following categories of participants: women and men; those who had close family relationships and those who did not; and those who had close friend relationships and those who did not. I compared responses on these measures for the objective, or behavioral, harassment indicators that at least 10% of respondents reported experiencing as well as participants' own perceptions of whether their experiences qualified as harassment. Table 3.2 presents results from *t*-tests of differences on the objective harassment indicators and on participants' perceptions of whether experiences qualified as various forms of harassment.

Tests for gender differences reveal that there were no statistically significant differences between men and women in their responses to the objective harassment indicators. There were also no significant differences between men's and women's *perceptions* of their experiences.

In the survey, participants were asked whether they would define any of their experiences as bullying, hostile, harassment, or sexual harassment. Based on findings in my prior research on younger workers (Uggen and Blackstone 2004), I hypothesized that men would be significantly less likely than women to define their experiences as harassment. Cultural realities, as well as stereotypes about gender, prescribe that women are more likely than men to be victimized. Men are also less likely than women to be socialized to think of themselves as potential victims and I therefore hypothesized that they would be less likely to perceive that they had been harassed. The finding in the present study that men and women did *not* differ in the likelihood of perceiving that they had been targets of workplace harassment suggests that the salience of gender as a part of workers' identities may differ based on a worker's age. Indeed, prior research on gender and aging processes generally (e.g., Meadows and Davidson 2006) and older workers specifically (e.g., McMullin and Berger 2006) suggests that older individuals may experience bullying or hostility from coworkers due to their age more than their gender.

The *t*-test results suggest that closeness with family and friends had some impact on harassment experiences but almost no impact on perceptions of workplace harassment experiences. Predictions from routine activities theory suggest that close family and friend relationships may serve a protective function; those with close ties may be perceived by harassers to be less vulnerable targets as they are more likely to have the resources to report and respond to harassment. Conversely, those with close ties might be *more* likely to *perceive* harassment as individuals with whom participants have close relationships can serve as "sounding boards" for participants, enabling them to examine their workplace experiences more critically than they might without others with whom to discuss them.

In terms of behavioral experiences, those who reported having close relationships with family were less likely than those lacking close family relationships to report being talked down to by bosses, less likely to report having their work contributions ignored, and less likely to report being isolated from important work activities. Those who reported having close relationships with friends were less likely than those lacking close friend relationships to report being isolated from important work activities and less likely to report experiencing unwanted questions about their private lives.

While closeness with family had no effect on perceptions of any kind of harassment, closeness with friends was significantly related to perceptions of sexual harassment. As shown in Table 3.2, those who had close friends were more likely than those who did not to *perceive* their experiences as sexual harassment. Of course, one should take into account that the findings reported in Table 3.2 are rather weak given the number of *t*-tests run.

Objective 3: How Do Older Workers Respond to Potentially Harassing Behaviors?

Table 3.3 presents findings from the questions asking older workers to whom they spoke about events they believed to be harassment.

Table 3.2 Percent reporting harassment behaviors and perceptions: independent sample *t*-tests of differences by gender, family relationships, and friend relationships

Behavioral indicators	Gender		Close w/Family		Close w/Friends	
	% Women	% Men	% Yes	% No	% Yes	% No
Verbal exchanges characterized by yelling or swearing	14.5	9.3	11.0	12.9	13.4	10.3
Talked down to by coworkers	21.7	18.6	18.5	25.8	23.5	17.2
Talked down to by bosses	19.6	23.3	17.8#	32.3	21.0	20.7
Offensive age-related joking about you	9.4	11.6	8.2	9.7	10.1	10.3
Offensive age-related joking about others	11.6	11.6	11.0	9.7	10.1	13.8
Offensive sexual joking about others	9.4	14.0	9.6	12.9	10.9	10.3
Work contributions ignored	25.4	23.3	21.2*	41.9	26.1	25.9
Isolated from important work activities	11.6	14.0	10.3*	25.8	9.2*	20.7
Left out of decisions that affect your work	23.9	18.6	21.9	25.8	21.9	25.8
Comments or behaviors that demean your age	13.8	9.3	11.6	12.9	14.3	10.3
Unwanted questions about your private life	10.9	9.3	8.9	16.1	7.6#	17.2
Perceptions of harassment						
Would you consider any of your experiences noted above to be…						
…Bullying?	13.0	14.0	13.7	12.9	16.0	8.6
…Hostile?	23.2	14.0	19.2	19.4	22.7	17.2
…Harassment?	16.7	16.3	15.8	16.1	19.3	12.1
…Sexual harassment?	4.3	9.3	4.8	6.5	7.6#	1.7

#p<.10; *p<.05; **p<.01

Sample sizes 138 for women and from 43 for men; 146 for close to family and 31 for not close to family; 119 for close to friends and 58 for not close to friends

Table 3.3 Persons older workers told about experiences of harassment (percentages)[a]

If yes to any of above, with whom did you speak about those events?	Co-worker	Boss	Family/Friend	Attny/Govt	Other	None
…Verbal exchanges characterized by yelling or swearing (n=24)	**45.8**	**33.3**	**8.3**	**0.0**	**4.2**	**37.5**
…Being talked down to by coworkers (n=38)	**50.0**	**26.3**	**23.7**	**0.0**	**5.3**	**31.6**
… Being talked down to by customers (n=15)	0.0	13.3	13.3	0.0	13.3	60.0
…Being talked down to by bosses (n=37)	**16.2**	**40.5**	**27.0**	**0.0**	**10.8**	**35.1**
…Offensive age-related joking about you (n=18)	**27.8**	**11.1**	**11.1**	**0.0**	**5.6**	**50.0**
…Offensive age-related joking about others (n=21)	**19.0**	**9.5**	**9.5**	**0.0**	**4.8**	**66.7**
…Offensive sexual joking about you (n=5)	40.0	20.0	0.0	0.0 %	0.0 %	60.0
…Offensive sexual joking about others (n=19)	**26.3**	**5.3**	**10.5**	**0.0**	**5.3**	**57.9**
…Having your work contributions ignored (n=46)	**15.2**	**34.8**	**19.1**	**0.0**	**4.3**	**47.8**
…Subtle or obvious threats to your safety (n=6)	16.7	83.3	0.0	0.0	16.7	16.7
…Bribes to do things that are wrong (n=3)	0.0	66.7	0.0	0.0	0.0	33.3
…Being hit, pushed or grabbed (n=5)	20.0	60.0	0.0	0.0	20.0	20.0
…Being isolated from important work activities (n=23)	**8.7**	**30.4**	**21.7**	**0.0**	**8.7**	**52.5**
…Being left out of decisions that affect your work (n=42)	**23.8**	**35.7**	**14.3**	**0.0**	**2.4**	**47.6**
…Comments or behaviors that demean your gender (n=10)	10.0	0.0	30.0	0.0	0.0	60.0
…Comments or behaviors that demean your age (n=23)	**21.7**	**26.1**	**17.4**	**0.0**	**0.0**	**43.5**
…Unwanted questions about your private life (n=19)	**10.5**	**26.3**	**21.1**	**5.3**	**10.5**	**45.0**
…Staring or invasion of your personal space (n=14)	21.4	21.4	28.6	0.0	7.1	35.7
…Unwanted touching (n=3)	0.0	100	0.0	0.0	0.0	0.0
…Sexually explicit pictures, posters, or other materials (n=2)	0.0	0.0	50	0.0	0.0	50
…Sexual assault by a coworker or boss (n=2)	0.0	0.0	0.0	0.0	0.0	100

[a]Because participants were asked to select as many answers as applied for each behavior reported, rows do not always total to 100 %. Bolded items are those for which at least 10 % of participants said they had the experience. In each row, the cell containing the largest font represents the most common response to each experience

Twenty-four participants experienced verbal exchanges characterized by yelling or swearing. Of these, 45.8% told a coworker; 33.3% told a boss; 8.3% told a family member or friend; 4.2% told someone other than those specifically asked about on the survey; and 37.5% told no one.

Overall, there are several notable patterns in Table 3.3. First, only one participant told an attorney or government agency about any experience. The single instance in which a participant told an attorney or government agency about his/her experience was in response to unwanted questions about his/her private life. Second, participants were likely to confront their harassers directly: the largest "who told" cell in the "talked down to by coworkers" row is coworkers and the largest "who told" cell in the "talked down to by bosses" is bosses. However, many participants told no one about their experiences of harassment.

Objective 4: What Do Older Workers Themselves Have to Say About Their Workplace Experiences?

As noted, I included three open-ended questions on the survey, asking participants whether they believed any of their harassment experiences had anything to do with their age or gender (and, if so, to explain) and to describe what they have most enjoyed about their work experiences since turning 62. In addition, many participants wrote unsolicited comments in the margins of the survey. While a small number of participants reported that their gender or age had something to do with their harassment experiences, far more participants wrote in the margins of their surveys, describing reasons for continuing work at their age. Most common reasons cited for working included financial need, maintaining social connections, and remaining mentally, physically, and/or emotionally functional.

Participants who added written remarks to the survey more typically wrote about positive, rather than negative, workplace experiences. Those who reported positive experiences said that working helped them maintain their independence; gave them a chance to socialize; provided them an opportunity both to learn and to share their knowledge; kept them busy; and made them feel useful and helpful. For example, one woman wrote, "I take pleasure to be with people. I love working. It keeps your mind alive, your body moving. I have a reason to get up in the morning." Another said, "My co-workers are fun to work with and I enjoy being able to earn some very much needed and appreciated money. I also appreciate the opportunity to learn new things." One woman wrote that she most enjoys, "Pride in my work, sense of accomplishment, making money, having health benefits, a purpose to my day, and being social." One man reported that he enjoyed "independence and control over my workplace and duties."

Participants who reported negative experiences said that working was sometimes difficult due to disability; that it sometimes felt demeaning to have to work; that their work environment was not enjoyable; and that they experienced ageism at work. For example, one woman wrote, "Because of my age, I was expected to remember way more than normal. Other people that did that job told me that it was a big guessing game." Another woman wrote, "Because I was older, the young girls thought I shouldn't be working, I should be at home waiting to die! They did everything they could to make my life there miserable and since the boss didn't try to stop it, I had to be quiet." One woman wrote,

> After about age 60–65 I began to notice that people would sometimes ignore me as though I had become invisible. This is also evident in the way that co-workers, supervision, prospective employers and others discount or ignore my ideas, opinions, views and work experience that may be extremely relevant or helpful. One is also patronized or talked down to in many situations where younger people think I have no value. I am not able to find work at the level I had before this age.

Conclusion

In this section, I summarize findings from my study, provide a few caveats, and consider what the findings suggest in terms of how older workers' experiences might be improved. The most common harassment behaviors included having one's work contributions ignored; being left out of decisions that affect one's work; and being talked down to by coworkers and bosses. All of these behaviors share themes of isolation and personal devaluation. The *lack* of difference between men's and women's perceptions of their experiences suggests that gender may be a less salient aspect of workplace identity for older workers than it is for younger workers. When sharing their experiences in their own words, as was the case for those participants who chose to write additional comments on their surveys, many reported positive workplace experiences. Finally, and perhaps most importantly, those who did experience harassing behaviors tended not to talk with others about them.

What do these findings suggest about how we might improve workplace experiences for older people? Even though older workers tend not to discuss their harassment experiences with others, coworkers do appear to be a potential source of support. Educating employers about the importance of nurturing positive coworker relationships may be one way to enhance older workers' experiences. Findings from this study also demonstrate the need to speak with older workers directly—both to hear more about the sorts of experiences they reported in the survey and to hear from them about how to improve their experiences. Qualitative research is needed in order to learn more from older workers themselves.

The present findings suggest common patterns among older workers' experiences. However, the sample from which these findings were drawn was not representative of all older workers. The small sample size, combined with the limited pool from which the sample was drawn (only participants from one state who happened to utilize the services of the worker placement/training program that assisted with data collection), limit the usefulness of this study for understanding the general nature of harassment of older workers. Nevertheless, this study does offer an important first step toward understanding and improving the workplace experiences of older people.

References

Acker, J. (1990). Hierarchies, jobs, bodies: A theory of gendered organizations. *Gender and Society, 4*, 139–158.

Allen, K. R., & Walker, A. J. (2006). Aging and gender in families: A very grand opening. In C. Toni & K. F. Slevin (Eds.), *Age matters: Realigning feminist thinking* (pp. 155–174). New York: Routledge.

Altschuler, J. (2004). Beyond money and survival: The meaning of paid work among older women. *International Journal of Aging & Human Development, 58*, 223–239.

American Association of Retired Persons (AARP). (2004). *Baby boomers envision retirement II*. Washington, DC: American Association of Retired Persons.

Appold, S. J. (2004). How much longer would men work if there were no employment dislocation? Estimates from cause-elimination work life tables. *Social Science Research, 33*, 660–680.

Barnett, R. C. (2005). Ageism and sexism in the workplace. *Generations, 29*, 25–30.

Bendick, M., Jr., Brown, L. E., & Wall, K. (1999). No foot in the door: An experimental study of employment discrimination against older workers. *Journal of Aging & Social Policy, 10*, 5–23.

Blackstone, A. (2003). Gender roles and society. In J. R. Miller, R. M. Lerner, & L. B. Schiamberg (Eds.), *Human ecology: An encyclopedia of children, families, communities, and environments*. Santa Barbara: ABC-Clio.

Brooke, L., & Taylor, P. (2005). Older workers and employment: Managing age relations. *Ageing & Society, 25*, 415–429.

Bumiller, K. (1988). *The civil rights society: The social construction of victims*. Baltimore: Johns Hopkins University Press.

Calasanti, T. (1999). Impact of work on older adults (book review). *Contemporary Sociology, 28*, 178–179.

Calasanti, T. (2002). Work and retirement in the 21st century: Integrating issues of diversity and globalization. *Ageing International, 27*, 3–20.

Calasanti, T., & Slevin, K. F. (2001). *Gender, social inequalities, and aging*. Walnut Creek: AltaMira Press.

Calasanti, T., & Slevin, K. F. (2006). *Age matters: Realigning feminist thinking*. New York: Routledge.

Cohen, L. E., & Felson, M. (1979). Social change and crime rate trends: A routine activity approach. *American Sociological Review, 44*, 588–608.

Connolly, C. (2002). The voice of the petitioner: The experiences of Gay and Lesbian parents in successful second-parent adoption proceedings. *Law and Society Review, 36*, 325–346.

De Coster, S., Estes, S. B., & Mueller, C. W. (1999). Routine activities and sexual harassment in the workplace. *Work and Occupations, 26*, 21–49.

Dennis, H., & Thomas, K. (2007). Ageism in the workplace. *Generations, 31*, 84–89.

DeSouza, E., & Solberg, J. (2004). Women's and men's reactions to man-to-man sexual harassment: Does the sexual orientation of the victim matter? *Sex Roles: A Journal of Research, 50*, 623–639.

Duncan, C. (2003). Assessing anti-ageism routes to older worker re-engagement. *Work, Employment and Society, 17*, 101–120.

Eichar, D. M., Norland, S., Brady, E. M., & Fortinsky, R. H. (1991). The job satisfaction of older workers. *Journal of Organizational Behavior, 12*, 609–620.

Ekerdt, D. J., Kosloski, K., & DeViney, S. (2000). The normative anticipation of retirement by older workers. *Research on Aging, 22*, 3–22.

Ewick, P., & Silbey, S. S. (1998). *The common place of law*. Chicago: University of Chicago Press.

Ewick, P., & Silbey, S. S. (2003). Narrating social structure: Stories of resistance to legal authority. *The American Journal of Sociology, 108*, 1328–1372.

Fairhurst, E. (2003). New identities of ageing: Perspectives on gender, age, and life after work. In A. Sara, D. Kate, & G. Jay (Eds.), *Gender and ageing: Changing roles and relationships* (pp. 31–46). Maidenhead: Open University Press.

Felstiner, W. L. F., Abel, R. L., & Sarat, A. (1980–81). The emergence and transformation of disputes: Naming, blaming, claiming.... *Law and Society Review, 15*, 631–654.

Fendrich, M., Woodword, P., & Richman, J. A. (2002). The structure of harassment and abuse in the workplace: A factorial comparison of two measures. *Violence and Victims, 17*, 491–505.
Fineran, S. (2002). Adolescents at work: Gender issues and sexual harassment. *Violence Against Women, 8*, 953–967.
Fleury-Steiner, B. (2003). Before or against the law? Citizens' legal beliefs and experiences as death penalty jurors. *Studies in Law, Politics, and Society, 27*, 115–137.
Ford, C. A., & Donis, F. J. (1996). The relationship between age and gender in workers' attitudes toward sexual harassment. *The Journal of Psychology, 130*, 627–633.
Giuffre, P. A., & Williams, C. L. (1994). Boundary lines: Labeling sexual harassment in restaurants. *Gender and Society, 8*, 378–401.
Hess, B. (1990). Gender and aging: The demographic parameters. *Generations, 14*, 12–15.
Hirsch, B. T., Macpherson, D. A., & Hardy, M. A. (2000). Occupational age structure and access for older workers. *Industrial & Labor Relations Review, 53*, 410–418.
Hoffman, E. A. (2003). Legal consciousness and dispute resolution: Different disputing behavior at two similar Taxicab companies. *Law & Social Inquiry, 28*, 691–716.
Hoffman, E. A. (2005). Dispute resolution in a worker cooperative: Formal procedures and procedural justice. *Law and Society Review, 39*, 51–82.
Hull, K. E. (2003). The cultural power of law and the cultural enactment of legality: The case of same-sex marriage. *Law & Social Inquiry, 28*, 629–657.
Kanter, R. M. (1977). *Men and women of the corporation*. New York: Basic Books.
King, N. (2006). The lengthening list of oppressions: Age relations and the feminist study of inequality. In C. Toni & K. F. Slevin (Eds.), *Age matters: Realigning feminist thinking* (pp. 47–74). New York: Routledge.
Knefel, A., Callender, M., & Cryant, C. D. (2004). Workplace as combat zone: Reconceptualizing occupational and organizational violence. *Deviant Behavior, 25*, 579–601.
Koonin, M., & Green, T. M. (2004). The emotionally abusive workplace. *Journal of Emotional Abuse, 4*, 71–79.
Kritzer, H. M., Vidmar, N., & Bogart, W. A. (1991). To confront or not to confront: Measuring claiming rates in discrimination grievances. *Law and Society Review, 25*, 875–887.
Levy, J. A. (1988). Intersections of gender and aging. *The Sociological Quarterly, 29*, 479–486.
MacKinnon, C. (1979). *Sexual harassment of working women: A case of sex discrimination*. New Haven: Yale University Press.
Macnicol, J. (2006). *Age discrimination: An historical and contemporary analysis*. Cambridge: Cambridge University Press.
Marshall, N. L. (2001). Health and illness issues facing an aging workforce in the new millennium. *Sociological Spectrum, 21*, 431–439.
Marshall, A. M. (2005). Idle rights: Employees' rights consciousness and the construction of sexual harassment. *Law and Society Review, 39*, 83–123.
Marshall, A. M., & Barclay, S. (2003). In their own words: How ordinary people construct the legal world. *Law & Social Inquiry, 28*, 617–628.
Marshall, V. W., Heinz, W. R., Kruger, H., & Verma, A. (Eds.). (2001). *Restructuring work and the life course*. Toronto: University of Toronto Press.
McLaughlin, H., Uggen, C., & Blackstone, A. (2008). Social class differences in workplace harassment during the transition to adulthood. *New Directions for Child and Adolescent Development, 119*, 85–98.
McMullin, J. A. (2000). Diversity and the state of sociological theory. *Gerontologist, 40*, 517–530.
McMullin, J. A., & Berger, E. D. (2006). Gendered ageism/age(ed) sexism: The case of unemployed older workers. In C. Toni & K. F. Slevin (Eds.), *Age matters: Realigning feminist thinking* (pp. 201–223). New York: Routledge.
McMullin, J. A., & Marshall, V. W. (2001). Ageism, age relations, and garment industry work in Montreal. *Gerontologist, 41*, 111–122.
Meadows, R., & Davidson, K. (2006). Maintaining manliness in later life: Hegemonic masculinities and emphasized femininities. In C. Toni & K. F. Slevin (Eds.), *Age matters: Realigning feminist thinking* (pp. 295–312). New York: Routledge.

Merry, S. E. (1990). *Getting justice and getting even: Legal consciousness among working class Americans*. Chicago: University of Chicago Press.
Minichiello, V., Browne, J., & Kendig, H. (2000). Perceptions and consequences of ageism: Views of older people. *Ageing and Society, 20*, 253–278.
Mor-Barak, M. E. (1995). The meaning of work for older adults seeking employment: The generativity factor. *International Journal of Aging & Human Development, 41*, 325–344.
Mortimer, J. T. (2003). *Working and growing up in America*. Cambridge, MA: Harvard University Press.
Mueller, C. W., De Coster, S., & Estes, S. B. (2001). Sexual harassment in the workplace: Unanticipated consequences of modern social control in organizations. *Work and Occupations, 28*, 411–446.
Nacelewicz, T. (2005). The graying of Maine. *Maine Sunday Telegram, 117*(40), 1–11.
National Institute for Occupational Safety and Health (NIOSH). (2004). *Worker health chartbook*. Cincinnati: NIOSH Publications Dissemination.
National Institute on Aging (NIA). (2007). *Growing older in America: The health and retirement study*. Bethesda: NIA.
Padavic, I., & Orcutt, J. D. (1997). Perceptions of sexual harassment in the Florida legal system: A comparison of dominance and spillover explanations. *Gender and Society, 11*, 682–698.
Pitt-Catsouphes, M. (2005, September 6). *Grant to study work options as boomers reach retirement*. Press release. Chestnut Hill: Boston College.
Richman, K. D. (2001). In times of need: Abused women's sources of support and changes in legal consciousness. *Studies in Law, Politics, and Society, 22*, 171–194.
Richman, J. A., Rospenda, K. M., Nawyn, S. J., Flaherty, J. A., Fendrich, M., Drum, M. L., & Johnson, T. P. (1999). Sexual harassment and generalized workplace abuse. *American Journal of Public Health, 89*, 358–363.
Rix, S. E. (2001). Restructuring work in an aging America: What role for public policy. In V. W. Marshall, W. R. Heinz, K. Helga, & V. Amil (Eds.), *Restructuring work and the life course* (pp. 375–396). Toronto: University of Toronto Press.
Rix, S. E. (2005). *Update on the older worker: 2004*. Washington, DC: American Association of Retired Persons Public Policy Institute.
Rospenda, K. M. (2002). Workplace harassment, services utilization, and drinking outcomes. *Journal of Occupational Health Psychology, 7*, 141–155.
Rospenda, K. M., & Richman, J. A. (2004). The factor structure of generalized workplace harassment. *Violence and Victims, 19*, 221–239.
Rospenda, K. M., Richman, J. A., & Nawyn, S. J. (1998). Doing power: The confluence of gender, race, and class in Contrapower sexual harassment. *Gender and Society, 12*, 40–60.
Rospenda, K. M., Richman, J. A., Wislar, J. S., & Flaherty, J. A. (2000). Chronicity of sexual harassment and generalized work-place abuse: Effects on drinking outcomes. *Addiction, 95*, 1805–1820.
Schaie, K. W., & Schooler, C. (Eds.). (1998). *Impact of work on older adults*. New York: Springer.
Siegenthaler, J. K., & Brenner, A. M. (2000). Flexible work schedules, older workers, and retirement. *Journal of Aging & Social Policy, 12*, 19–34.
Singh, G., & Verma, A. (2003). Work history and later-life labor force participation: Evidence from a larger telecommunications firm. *Industrial & Labor Relations Review, 56*, 699–715.
Smith, J. W., & Calasanti, T. (2005). The influences of gender, race, and ethnicity on workplace experiences of institutional and social isolation: An exploratory study of university faculty. *Sociological Spectrum, 25*, 307–334.
Sterns, H. L., & Miklos, S. M. (1995). The aging worker in a changing environment: Organizational and individual issues. *Journal of Vocational Behavior, 47*, 248–268.
Sterns, H. L., Barrett, G. V., Czaja, S. J., & Barr, J. K. (1994). Issues in work and aging. *Journal of Applied Gerontology, 13*, 7–19.
Taylor, P., & Walker, A. (1998). Employers and older workers: Attitudes and employment practices. *Ageing & Society, 18*, 641–658.
Texeira, M. Y. (2002). 'Who protects and serves me?' A case study of sexual harassment of African American women in one U.S. law enforcement agency. *Gender and Society, 16*, 524–545.

Tougas, F., Lagace, M., De La Sablonniere, R., & Kocum, L. (2004). A new approach to the link between identity and relative deprivation in the perspective of ageism and retirement. *International Journal of Aging & Human Development, 59*, 1–23.

Uggen, C., & Blackstone, A. (2004). Sexual harassment as a gendered expression of power. *American Sociological Review, 69*, 64–92.

U.S. Bureau of the Census. (2006). *Statistical abstract of the United States* (125th ed.). Washington, DC: U.S. Government Printing Office.

U.S. Equal Employment Opportunity Commission (EEOC). (2006). *Discrimination by type: Facts and guidance*. Retrieved January 30, 2006, from http://www.eeoc.gov/

Wakefield, S., & Uggen, C. (2004). The declining significance of race in federal civil rights law: The social structure of employment discrimination claims. *Sociological Inquiry, 74*, 128–157.

Wayne, J. H. (2000). Disentangling the power bases of sexual harassment: Comparing gender, age and position power. *Journal of Vocational Behavior, 57*, 301–325.

Welsh, S. (1999). Gender and sexual harassment. *Annual Review of Sociology, 25*, 169–190.

Welsh, S., Carr, J., MacQuarrie, B., & Huntley, A. (2006). 'I'm not thinking of it as sexual harassment': Understanding harassment across race and citizenship. *Gender and Society, 20*, 87–107.

Chapter 4
What Do Newspapers Say About Older Adults in the Workforce?

Mebane Powell

Introduction

Mass media reach large audiences and are comprised of various traditional and digital forms including print (e.g., books, newspapers), audio recordings and reproductions, broadcasting (e.g., radio and television) and film (e.g., cinema, DVDs). Historically, one of the most influential forms of mass media is newspapers. Various studies have found that U.S. newspapers are read by nearly three in four adults (Scarborough Research 2009; Urban 1999).

Newspapers can present issues in biased ways by favoring select social and political agendas or groups. The Madrid International Plan of Action on Ageing 2002 (MIPAA), the outcome document of the United Nations Second World Assembly on Ageing held in Madrid, Spain, in 2002, represents an international aging agenda intended to integrate the rights and needs of older people into national and international economic and social development policies (Global Action on Aging 2007). MIPAA notes the need to promote positive images of aging by encouraging the media and the private and public sectors to avoid ageism in the workforce and to present positive images of older persons (United Nations 2003).

Newspapers reach a large audience and play an important role in promoting or inhibiting public views of active and dignified aging. By examining newspaper articles, it is possible to study how portrayals of older adults in the workforce reflect positive or negative public perceptions of older adults. These public perceptions influence how individuals, including older adults themselves, think about aging and productivity; the treatment of older workers; and the formation and implementation of public and private policies and programs that affect older workers.

M. Powell, M.S.W. (✉)
Graduate School of Social Service, Fordham University, New York, NY 10023, USA

NYC Department for the Aging, Research Division, New York, NY, USA
e-mail: mpowell@fordham.edu

P. Brownell and J.J. Kelly (eds.), *Ageism and Mistreatment of Older Workers: Current Reality, Future Solutions*, DOI 10.1007/978-94-007-5521-5_4,

Framing as a News Reporting Strategy

Media framing is the process by which a form of mass media "constructs and defines a social or political issue for its audience" (Nelson et al. 1997, p. 221). Framing also has been defined as a reporting technique used by the news media to tell stories that reflect implicit social biases and values (Scheufele and Scheufele 2010). News framing at the social level is concerned with how dominant frames are established to influence public attitudes or policy outcomes. Frames that journalists infuse into their news stories can also influence how readers make sense of the world around them and make judgments about issues. Research by Bonnesen and Burgess (2004) found that newspaper articles reflected ageist attitudes, and included the widespread use of negative terms such as 'senior moment'. One investigation that applied framing and discourse analysis to newspaper articles in Ireland found "five distinct identity types . . . constructing older adults as 'victims'; 'frail, infirm and vulnerable'; 'radicalized citizens'; 'deserving old' and 'undeserving old'" (Fealy et al. 2011, p. 1). Although it is not the intent of this chapter to engage in frame analysis (for a comprehensive overview of this topic see D'Angelo and Kuypers 2010), it is a useful approach to categorizing news reports of older adults in the workforce and identifying sub-categories or themes that emerge from these reports.

In this chapter, the author presents the results of a qualitative content analysis of United States newspaper articles that appeared in print from February 1, 2010 to January 31, 2011, and reported on older persons in the workforce, as well as retirees. The purpose of this analysis was to determine if these newspaper articles portrayed older workers in a positive or negative way, and to identify the themes associated with these portrayals. In addition, this analysis sought to document how newspaper stories illuminated the realities of the workplace for older workers on issues such as age-biased hiring practices, the mistreatment of older workers, and the impact on older workers of changing economic conditions and retirement practices, as well as the positive effects of an older workforce on older workers themselves and on the companies that employ them.

Method

Sample

Proquest National Newspaper Premier, a newspaper database bundle option provided by Proquest National Newspaper, was accessed through Fordham University's electronic database and index library services (http://www.library.fordham.edu/database/newspaper.html) and was used to retrieve newspaper articles related to older adults and work. Information obtained from the Audit Bureau of Circulation (2011) showed that based on the most recent 6-month daily circulation average (April 2011–September 2011) Proquest contained 14 of the top 25 circulating

Table 4.1 Comparison of top 25 newspapers in US and Proquest National Newspaper premier newspapers

Audit bureau of circulations data April–September 2011	Ranking based on average daily circulation	Included in Proquest National Newspaper premier[a]	Average daily
Wall Street Journal	1	Yes	2,096,169
USA Today	2	Yes	1,784,242
The New York Times	3	Yes	1,150,589
New York Daily News	4	Yes	605,677
Los Angeles Times	5	Yes	572,998
Mercury News	6	No	527,568
New York Post	7	No	512,067
Washington Post	8	Yes	507,465
Chicago Tribune	9	Yes	425,370
The Dallas Morning News	10	No	409,642
Newsday	11	No	404,542
Chicago Sun-Times	12	No	389,353
Houston Chronicle	13	Yes	369,710
The Denver Post	14	Yes	353,115
Philadelphia Inquirer	15	Yes	331,134
Minneapolts Star Tribune	16	Yes	298,147
Arizona Republic of Phoenix	17	No	292,838
Organge County Register	18	Yes	270,809
The Plain Dealer of Cleveland	19	No	243,299
The Seattle Times	20	No	242,814
The Oregonian of Portland	21	No	242,784
St. Petersburg Times	22	Yes	240,024
The Detroit Free Press	23	No	234,579
San Francisco Chronicle	24	Yes	220,515
San Diego Union-Tribune	25	(Abstracts through December 2010)	219,347
Las Vegas Review-Journal	26	Yes	213,078
The Boston Globe	28	Yes	205,939
The Sacramento Bee	29	Yes	199,921
St. Louis Post-Dispatch	31	Yes	191,631
The Atlanta Journal-Constitution	38	Yes	173,884
Pitts burgh Post-Gazette	39	Yes	173,160
South Florida Sun-Sentinel	45	Yes	147,860
San Antonia Express-News	53	Yes	137,514
Detroit News	59	Yes	128,571
Seattle Post-Intelligencer	No data	Yes	No data
The Christian Science Monitor	No data	Yes	No data

[a]Other data sources included: *Wall Street Journal Online*; WSJ: the Magazine from the Wall Street Journal: *New York Times Book Review*; *New York Times Magazine*; Seattlepi.com

newspapers in the U.S., including *The New York Times*, *Wall Street Journal* and *USA Today*. Proquest also included 11 newspapers whose rankings fell outside the top 25, and also five other sources Proquest categorized as newspaper sources (e.g., *Wall Street Journal Online*) (see Table 4.1).

In order to find newspaper articles published between February 1, 2010 and January 31, 2011 related to the workforce of older adults, Proquest was searched using each of the following phrases as subject terms: "older worker" (N=111); "age discrimination" (N=72); "retirement" (N=2,704); "work environment" (N=126); and "volunteer" (N=1,513). Due to the large number of results for "retirement" and "volunteer", these subject areas were narrowed to select only articles that focused on "retirement" and "volunteering", with any of the following as subjects: older people, older workers, senior citizens, elderly people, and older adults. This resulted in 105 articles returned for "retirement" and 50 for "volunteer". As a result of these selections, a total of 464 articles were retrieved, 46 of which were identified as exact duplicates. Of the remaining 418 newspaper articles published between February 1, 2010 and January 31, 2011, a review for content that referred either directly or indirectly to older adults in the workforce was conducted to warrant inclusion in the study, resulting in a final sample of 97 articles. Each article was independently categorized by the author as reflecting a positive or negative theme of older adults in the workforce and retirees.

Themes are often used to measure the presence of frames. According to Levin (2005), "The reason themes [are] taken as a measure of the presence of frames [is] the difficulty of finding a completely developed frame in a single press release. [Frames] are built across a series of news media articles, and not all elements are present in any single article" (p. 89) (as cited by D'Angelo and Kuypers 2010, p. 319). In this chapter, themes were grouped into four frames. Frames were developed based on the dichotomy of negative/positive themes indicating negative/positive frames. These frames were then further refined by grouping them into the dichotomy of macro versus micro frames. Frames included: (1) older adults are portrayed as a problem in the workforce (negative macro frame); (2) older adults reported negative experiences in the workforce (negative micro frame); (3) older adults reported positive experiences in the workforce (positive micro frame); and (4) older workers were portrayed as having a positive effect on the workforce (positive macro frame).

Findings

Older adults are portrayed as a problem in the workforce (negative macro frame).

There were a total of 23 articles that met the criteria for this category. These in turn yielded the following themes or sub-categories: (1) unwilling to leave the workforce (n=9); (2) lack of respect for younger managers and bosses (n=2); (3) high cost to society of benefits and entitlements (n=10); and (4) older workers retaliate by suing their employers (n=2).

Unwilling to Leave the Workforce

Norris (2011)noted that "thirty years ago, one in seven jobs in the United States was held by a person who was 55 or older. Today the proportion is one in five" (p. B3). Other reporters echoed this view (Abate 2010a; Beyer 2010; Cauchon 2010;

Denning 2010; Enforce limits on noise from motorcycles 2011; Julian 2010; Norris 2010; Pugh 2010). One article stated that government jobs are hard to get as "[d]ue to the recession, fewer employees are retiring early . . ." (Beyer 2010, p. 1). Carl Van Horn, director of the Heldrich Center of Workforce Development at Rutgers recognized that older adults in the workforce are remaining because not everyone can afford to retire and added that this "can indirectly hurt younger workers" as "there are only so many jobs to go around" (Cauchon 2010, p. A1). This complaint about displacement of younger workers was echoed by an article reporting that July 2010 was the worst month since July 1948 for 16–24 year olds obtaining employment due to older adults in the workforce not retiring (Denning 2010, p. C20).

Lack of Respect for Younger Managers and Bosses

Another way newspaper articles reported older adults in the workforce as a problem for younger workers was their lack of acceptance of reporting to a younger manager. For example, Stafford (2010) described an older worker who complained about a younger female manager by saying, . "I've forgotten more than that missy knows." This worker wanted to quit because a younger woman had been promoted to be his boss (p. F1). Pritscher (2010) reported on a 26 year-old hiring manager who worried about how older workers reacted to him: "Older employees have made comments occasionally, like, 'How did you get to be the one that hires us?'" (p. BU9).

High Cost to Society of Benefits and Entitlements

Newspaper articles also portrayed older adults in the workforce as a problem by illustrating the high cost to society for their benefits and entitlements (Murray 2010). For example, one California newspaper reported,

> Public employee retirement benefits in California are the most generous of any in the country. They've gone from just 40 percent of pay for the average 30-year state employee when the retirement system was established in 1932 to close to 75 percent of pay today. Many local government workers earn more the day they retire than when they worked. Given earlier retirement ages -- 55 for most public employees and 50 for firefighters, police and prison guards -- and longer life spans, many retired state workers will pick up more retirement checks in their lifetimes than paychecks. Such high costs are unsustainable, economically and politically (Clock ticks on bloated pensions 2011, p. A6).

This article highlighted the fact that governments are unable to sustain pension programs at current levels due to the number of people retiring, coupled with longer life spans. One article suggested the need to "cut or at least slow the growth of retirement benefits in order to deal with runaway deficits" (What's the big idea? 2011, p. B2), whereas other articles noted that the growing number of older workers who will be collecting Social Security and other benefits presents a problem (Klein 2010; Leland 2010; Murray 2010; Ohlemacher 2010a, b; Sestak 2010).

Older Workers Retaliate by Suing Their Employers

Two articles noted that older workers are more willing to sue (e.g., Pender 2010). For example, Pender reported an attorney as saying "[Older workers'] losses are so substantial, and the risk of making a claim are so small, there is much more to be gained than lost by proceeding with a claim against an employer" (p. D1).

Older Adults Experienced and Reported Problems in the Workforce (Negative Micro Frame)

A total of 45 news articles reported on older adults who identified problems they experienced in the workplace due to their age. Themes that emerged from the content analysis of these news stories included: (1) when searching for a job, older workers should hide or downplay their age (n = 12); (2) experienced mistreatment at work (n = 28); and (3) retirement plans altered (n = 5).

When Searching for a Job, Older Workers Should Hide or Downplay Their Age

These newspaper articles often provided tips to older workers for writing resumes or preparing for job interviews. The underlying theme of this advice was that society is ageist and older workers had better be savvy about hiding or downplaying their age. This was supported by a recent survey which found that "84% of managers say employers would hesitate to hire a qualified but visibly older candidate" (Smiley 2011, p. A8).

Several articles encouraged older adult workers to include only the last 10 years of employment and to omit the dates they obtained their degree (Rich 2010), suggesting that advanced years prejudiced the chances of older adults to obtain employment. The authors of these articles believed that older job applicants were less likely to be called for interviews if prospective employers knew their age. Other advice also reflected ageist attitudes in the workforce. For example, authors advised older job interviewees to highlight their ability to work with younger individuals and willingness to be managed by, and report to, a younger supervisor; to reinvent themselves; to show they have more energy; and to show they know how to use the latest technology (Abate 2010a; Bonnie 2010; Giegerich 2010; Grantham 2010a; Rich 2010).

The disadvantages of being older when applying for jobs were described by many older applicants. Typical remarks included, "You get caught in the age crunch. A lot of people say age doesn't matter, but it does to employers" (Chapman 2010a,

p. A10). "The dates in the resume give away your age, so people don't even ask to see you" (Hsu 2010, p. B).

Several articles reported that older job applicants employed strategies to look younger (e.g., Bonnie 2010; Felten 2010). One such female applicant told the reporter, "I would do it all – Botox, lasers, everything – if I could afford it. If it meant getting hired, I would do whatever I could to stop time." Another applicant, an older male, had eyelid surgery to make himself look younger. As a result, he believed "his more refreshed image has provided a boost" (Bonnie 2010, p. 1.1).

Experienced Mistreatment at Work

As one newspaper article noted, "More than 63% of workers age 45 and older have either faced or observed age discrimination in the workforce" (Seniors don't just take, but they also give back 2010, p. A8). Other articles reported age-discrimination as the reason for not being hired or promoted, or for being fired (Chapman 2010a, b; Garone 2010; Lindgren 2010; Rampell 2010; Semuels 2010a, b).

Articles in this category included those reporting on suits against police departments (Kever 2010); repeat age discrimination offenders (Abelson 2010a, b; Dolnick 2011); and age discrimination suits against large organizations such as AT & T (Schultz 2010), Google Inc. (Dolan 2010; Efrati and Koppel 2010; Funt 2010), the FBI (Layton 2010), CBS (Bray 2010; Egelko 2010, 2011), and the NBA (Dillman 2010).

Dolan (2010) reported on a manager from Google who said his "colleagues refer to him as an 'old fuddy-duddy' and an 'old man'" and that a "high-level manager told him he was 'fuzzy,' 'sluggish,' 'lethargic' and did not 'display a sense of urgency'" (p. B4). Another article reported on an age-bias suit filed by the federal Equal Employment Opportunity Commission (EEOC) against a company for its practice of requiring partners who turned 70 to relinquish their ownership interests in the firm (Sachdev 2010, p. 16).

Although these and other cases (Alexander 2010; Wakin 2010) went to trial, many of the cases filed under the Age Discrimination in Employment Act (ADEA) were dismissed entirely or had parts of the lawsuit dismissed for lack of evidence (City/Allegany County 2010; Egelko 2010). Some lawsuits were settled (Egelko 2011; O'Reiley 2010; Pabst 2010; Sixel 2011). Such was the case in an age discrimination lawsuit filed by a 56 year old waitress in an adult entertainment club. She claimed that "male managers called her 'old' and said she showed signs of memory loss" (Sixel 2011, p. D3). In all these cases, plaintiffs face an often insurmountable hurdle: the high standard of burden of proof that age discrimination took place. This is reflected in the paucity of ADEA lawsuits in which the plaintiff was successful, and the many lawsuits that are dismissed or in which the plaintiff is forced to settle. (See Chap. 9 for a discussion of legal and legislative issues in older adult protection in the workplace.)

Retirement Plans Altered

Articles in this category reported on the downward turn of the economy and its impact on older workers (Dugas 2010; Farrell 2010; Lee 2010). One reporter noted that "the longer it takes for the economy to recover, the less money [older workers] will have to spend in retirement" (Another threat to economy 2010, p. 1). In addition, older workers may need to retire early before pensions, benefits and layoff packages are cut due to the poor economy. Some older workers may need to continue working past their planned retirement age (Another threat to economy 2010; Farrell 2010; Grantham 2010b).

One reporter described the new realities of retirement:

> The triple-punch of a severe recession and bear markets on Wall Street and in real estate has continued to rewrite the retirement plans of [older workers]. Though two of those obstacles have eased lately - with the recession generally considered to have ended in mid-2009 and the stock market rising roughly 60 percent since its early 2009 downturn - their effects linger. Many people have been forced into early retirement because of staff downsizing, while others, nervous after the recent economic events, have put their plans for leaving the workforce on hold (Grantham 2010b, p. D1).

Another reporter noted that "older workers, who typically fared better than their younger counterparts in recessions, have been hit just as hard by layoffs this time around….older workers have now found themselves at the back of the line to return to the work force" (Another threat to economy 2010, p. 1).

Older Adults Reported Positive Experiences in the Workforce (Positive Microframe)

A total of 20 news stories were identified in which older adults themselves reported positive experiences in the workforce. Themes that emerged from a content analysis of news reports that fell within this frame included: (1) older adults find personal benefit in work (n = 10); and (2) older workers report success in the workforce (n = 10).

Older Adults Find Personal Benefit in Work

Many older workers continued to work into their later years because they wanted to keep their minds sharp; others were not ready to retire fully for other reasons (Abelson 2010a; Axtman 2010; Miller 2010; Porter 2010; Steffy 2010; Vanderkam 2011; Ward 2010). Abelson (2010a) reported the results of interviews with several older workers, also past retirement age: "It's [the job] what keeps me going, you know"; "I just don't like hanging around, I want to meet people and keep busy";

"It's [work] very important mentally and physically. It gives you a great incentive to live a better life" (p. A1). Fernandez (2010) reported on the longest serving doorman of a union in New York City who, at 83, said "I love[s] it (the job)" and "[it]is never late" (p. A22). Other articles reported older workers who flaunted their age, gray hair and lack of Botox treatments as an asset (Schmich 2010). One reporter argued that staying in the workforce allowed older workers to retain health care coverage for themselves and their families, and that working part-time or flexible schedules allowed older workers to care for family members (Trumbull 2010).

Older Workers Report Success in the Workforce

Older Adults Create Their Own Employment

"Studies show that it is more difficult for older job seekers to enter or re-enter the workforce" (Abate 2010b, p. D2). One strategy used by older persons seeking work is to create their own jobs. For example, Greenhouse (2010) reported that "more than five million Americans age 55 or older run their own businesses or are otherwise self-employed, with a 52% increase from 2000–2007 of people ages 55-64 being self-employed" (p. F1). In addition to being self-employed, older adults seeking work are also creating new innovative models to ensure their skills are used or they are hired. As reported by Fishman (2010), "Sarasota, Florida, is flush with new not-for-profit organizations because retirees there crave new projects and start charities" (Fishman 2010).

Seizing Opportunities to Remain in the Workforce

Other articles (Goodman 2011; Pope 2010; Seltz 2010; Wolf 2011) indicated a trend of older adults entering "encore careers" that "match older workers who can't or don't want to retire with public service jobs that benefit society" (Wolf 2011, p. A1) or with volunteer work.

Margaret Harrison "adapted skills acquired in the corporate world to help low- and middle-income families buy homes" (Lotan 2010, p. G6). As Ms. Harrison reported, "You can't retire at 60 and just stop. . . I need something to keep my brain active" (Lotan 2010, p. G6). Paul Corcoran, 78, worked with children and young adults, assisting them in writing college applications, finding work, moving, and learning to read and write (Abraham 2010). In other areas of the country older adult volunteers help to fill gaps in services due to budget cuts and hiring freezes (Simmons 2010). Harold Adair, 71, "has proven his worth for the past 19 years by pushing buttons to let people through the controlled gates of the jail, taking telephone calls, acting as a receptionist, a typist and an all-around helper" (Simmons 2010, p. B1). Anne Sims, 78, volunteers by helping a little boy read. "[Volunteering] is something that means a lot to me," (Staples 2010, p. B1). Trudy Dickenson, 84, is known not only as a volunteer but as a mentor who continues to volunteer at a school because she enjoys the work and because "everybody is so welcoming," (Staples 2010, p. B1).

Although these articles do not represent a policy change in the workforce to support older workers, they do illustrate society's willingness to harness the value of older workers' experiences.

Older Workers Were Portrayed as Having a Positive Effect on the Workforce (Positive Macro Frame)

A total of nine articles were identified that reported older adults as enhancing the workforce environment. Themes or sub-categories that emerged from the content analysis within this frame included ways that older adults in the workforce are seen as valuable and include: (1) older workers are seen as a benefit to the company bottom line, rather than a deficit (n=4) and; (2) workers are portrayed as having vast knowledge (n=5).

Older Workers Are Seen as a Benefit to the Company Bottom Line, Rather Than a Deficit

Kaslow and Tate (2011), in an article published outside the timeframe of this analysis, illustrated this theme. The article was entitled: "*US economy's hidden asset: older workers; Far from being a drag on the economy, so-called gray labor will be key to America's competitiveness in coming years. Mature workers can bring major productivity gains to US businesses - if we can make changes to better tap their talent.*" The authors continued, "despite old perceptions that aging Americans drag on the economy, new realities demonstrate that properly primed, mature workers can be a key element of the talent pipeline so critical to the nation's competitiveness" (p. 1). Several articles reported on companies implementing organizational changes that have been recognized as models for hiring and retaining older workers and, in some instances, increasing workforce productivity as a result (Gallagher 2010; Hailey 2010; O'Connell 2010; Turner 2011).

Hiring Practices

In an article on older workers returning to work, (Hailey 2010) it was noted that "... some [businesses] increasingly value older workers for their experience and reliability . . . and have started modifying their hiring approaches to adjust to the needs of older workers" (p. O2). According to Hailey, "some companies, such as pharmacy chain CVS Caremark Corp. of Woonsocket, R.I., offer so-called snowbird programs that allow northern employees to transfer to branches in warmer places during winter. Others let employees reduce hours instead of retiring" (p. O2).

Older Workers Inform the Organization

Gallagher (2010), in an article about BMW, noted that they were "faced with an aging work force and declining productivity" and decided to respond to this issue by conducting an experiment "that has been hailed as a prototype for how to keep an aging workforce happy and productive" (p. 2A1). "By allowing workers to make whatever changes they felt were needed to keep productivity high, workers came up with 70 improvements . . . such as larger font on computer monitors, floors that were easier on the knees, flexible work hours, tools with larger handles" all of which resulted in a 7% increase in productivity in a year (Gallagher 2010, p. 2A1).

Workers Are Portrayed as Having Vast Knowledge

Five authors cited the benefits to employers of hiring and retaining older workers. Foremost among these benefits is that older workers often have vast experience and knowledge of how the company operates in relation to its historical performance and organizational environment (Channick 2010; Dickinson 2010; Leland 2010; Meyer 2010; Sulzberger 2010). "The older ones [workers] came to us after being family caregivers, so they understood the stresses that families were under" (Leland 2010, p. A14), and "Companies today do not and cannot afford to train workers [and] want to hire someone who can step in and do the job from day one" (Dickinson 2010, p. 2).

Older workers' ability to give advice and serve as mentors are cited as benefits by Leland (2010). One of Leland's interviewees stated, "If you look at older women as an asset to train younger workers, they can be valuable" (p. A14). An article in the *Houston Chronicle* reported the story of Sally Gordon, 101 years old, who was honored for being "America's Outstanding Older Worker for 2010". When asked if she had plans to retire, Sally said she did not, but went on to advise young people to "learn new techniques, listen and be kind to co-workers" (101-year-old Working Woman Best in U.S. 2010, p. 2).

Discussion

In this analysis of U.S. newspaper articles over a period of one year, 70% (n=68) of the articles fit primarily into the negative macro/micro frames. These included: (1) older adults are portrayed as a problem in the workforce (negative macro frame); and (2) older adults experienced and reported problems in the workforce (negative micro frame). The remaining 30% of articles (n=29) reflected positive experiences, primarily from the perspectives of older persons (positive micro frame), rather than from the portrayal of older adults as having a positive effect on the workforce environment (positive macro frame).

Results from the present study of newspaper articles support the findings of other researchers of bias towards older adults that is largely negative in nature (Bonnesen and Burgess 2004; Fealy et al. 2011; Murphy 2004; Rudman and Molke 2009). Thus, it may be expected that employers will favor younger people entering or in the workforce. Other indicators in newspapers that old age is portrayed in a negative light include articles that focus on beauty products to reduce wrinkles, cover gray hair, and ways to make one look younger (Fealy et al. 2011; Murphy 2004). The finding of primarily negative with some positive themes of ageism found in the current study supports themes found and noted in prior work by Palmore (1999), Nelson (2002) and the International Longevity Center (Dahmen and Cozma 2006). Palmore (1999) identified negative ways older adults are portrayed, which included: (1) illness; (2) impotency or lack of sexual activity; (3) ugliness or unattractiveness; (4) mental decline; (5) mental illness that is common and untreatable; (6) uselessness or inability to work due to disabilities; (7) isolation or lack of social networks; (8) poverty or wealth; and (9) being depressed. Nelson (2002) noted that the promotion of youth attributes, coupled with the cultural frame of aging stereotypes, indicates that older adults are negatively portrayed in the media by not having those qualities. The International Longevity Center (Media Takes On Aging, 2006) reported that a negative experience of older adults in the workforce is that of being forced to retire or having benefits altered or eliminated.

This preponderance of negative portrayals of older workers and negative experiences reflects the influence of pervasive ageist stereotypes. These stereotypes lead to age discrimination in the workplace. This was confirmed by AARP (2001) in a survey to "assess public awareness of age discrimination in the workforce" (p. 2). AARP researchers found that close to two-fifths (38%) of Utah residents aged 40 and older were aware of age discrimination in the workforce, either from first or second hand experience, or from reading or hearing about ageism in the news.

Negative stereotypes of older workers are further fueled by newspaper stories that focus on intergenerational conflict. This conflict is supported by newspaper stories and societal beliefs that older adults in the workforce benefit at the expense of younger individuals, that they are "benefit and entitlement rich", and that they harm younger workers by selfishly refusing to leave the workforce to make way for younger workers entering the job market. These views also encourage younger individuals to adopt negative ageist stereotypes and to promote workforce policies that encourage employers to hire younger workers. In many cases, older adults in the workforce, like their younger counterparts, have been hit hard by the economic recession. They stay in the workforce in an attempt to recoup losses to their retirement plans. The ability to engage in paid work longer to maintain financial security is crucial to the quality of life of both younger and older workers.

As stated by Carl E. Van Horn, professor of public policy and director of the John J. Heldrich Center for Workforce Development at Rutgers University, "Older workers who lose their jobs could pose a policy problem if they lose their ability to be self-sufficient" (Rich 2010, p. A11). According to a recent U.S. Labor Department report, "Of the 14.9 million unemployed, more than 2.2 million are 55 or older. Nearly half of them have been unemployed 6 months or longer. The unemployment rate in the group – 7.3% – is at a record, more than double what it was at the beginning of the latest recession" (Rich 2010, p. A11). However, according to the newspaper articles included in the present study, many employers value older workers and want to retain them in the workforce. Retaining older workers is a benefit to society because companies can count on workers who stay in the workforce to contribute longer to benefits and pension programs, thus costing younger workers less money in the long-term (Davidson 2010). In addition, as Vanderkam (2011) reported, there is "a recognition that people both need and want to be part of the workforce longer . . . and that seniors with incomes feel more secure and spend more in a way that generally boosts the economy" (p. A7).

Many older adults seek identity through their work and adherence to a work ethic. This identity, in turn, is an important determinant of the quality of life of older workers. According to a study of older workers conducted by the Pew Research Center, workers over the age of 30 cited work ethic as a "badge of their age groups' identity" at higher rates than younger workers (Shapira 2010, p. B1). Older adults' work ethic, coupled with their difficulty in remaining in the workforce, may be why many older adults are seizing the opportunity to become volunteers or engage in "encore careers". These encore careers and volunteer opportunities are proof of positive and productive aging; they are also strategies older workers use to remain active in the workforce in spite of barriers to their acceptance. Thus, older adults in the workforce are communicating that they are willing and able to work when given the opportunity. It is yet to be seen if the United States will embrace the work ethic of older adults and create policies that encourage the hiring and retention of older workers. The organizations cited in the present study that are models for implementing such policies represent one set of resources in beginning this process.

Findings from the present study also have implications for the implementation of an international agenda such as MIPAA (United Nations 2003) to promote positive aging. As stated in MIPAA (Priority area III, Issue 4: Images of aging, pp. 44–45), positive media coverage of older workers is important to promote images that highlight the wisdom, strengths, contributions, courage and resourcefulness of older women and men, including older persons with disabilities. It is also important for the media to move beyond portrayal of stereotypes and to illuminate the full diversity of older persons. The media are harbingers of change and can be guiding factors in fostering the role of older persons in society.

Non-governmental organizations (NGOs) can also assist in promoting a positive view of aging in the media, including incentivizing media focus on older people and productive aging. Most recently, the Gerontological Society of America (GSA) and New America Media (NAM) initiated a journalism fellowship program funded by the MetLife Foundation to encourage journalists in raising public awareness about issues of aging. IN the GSA program, journalism fellowships are awarded to 16 journalists annually who focus on aspects of aging and participate in forums that bring to the fore the latest findings in research and issues facing older adults.

One fellowship was awarded to a journalist, Kerry Hannon, author of a series of articles on work and retirement of the boomer generation in the United States (Gerontology News 2011). Innovative programs like this represent a proactive stance that NGOs can take to remedy the age-negative media coverage biases of older adults by highlighting active and productive aging. By providing journalists with stories that highlight positive aspects of aging and are informed by current research data, NGOs can become active players in informing the public about positive attributes of older workers and retirees.

Conclusion

Ageist attitudes are pervasive in American society. Thus, it is not surprising that negative portrayals of older workers predominate in U.S. newspaper articles, and that these articles often report on problems older adults experience in the workplace that are due to their age. Mass media, such as newspapers, reflect public attitudes toward older people. But as recognized in the MIPAA (UN 2003), mass media also shape public attitudes. To the extent that newspapers and other forms of mass media portray older persons as capable and productive workers, public attitudes will shift toward more realistic and positive views of older persons. In the long run, this will enhance the self-esteem of older workers and will lead to greater social and occupational choices for them. At the same time, it will enhance the performance of individual companies and of the economy as a whole, as the skills and experience of older workers are utilized. The end result will be an economic and moral enrichment of society itself.

References

101-year-old working woman best in U.S. (2010, August 11). *Houston Chronicle*, p. A2. Retrieved from Proquest National Newspaper Premier.

Abate, T. (2010a, July 14). Seniors delay retiring, making it tough for teens to find jobs. *San Francisco Chronicle*, p. A1. Retrieved from Proquest National Newspaper Premier.

Abate, T. (2010b, July 18). Get to work. *San Francisco Chronicle*, p. D2. Retrieved from Proquest National Newspaper Premier.

Abelson, J. (2010a, May 22). They're just not the retiring type. *Boston Globe*, p. A1. Retrieved from Proquest National Newspaper Premier.

Abelson, J. (2010b, June 24). Fired worker sues BJ's, claims age bias. *Boston Globe*, p. B7. Retrieved from Proquest National Newspaper Premier.

Abraham, Y. (2010, November 25). A show of good faith. *Boston Globe*, p. B1. Retrieved from Proquest National Newspaper Premier.

Alexander, S. (2010, August 26). Judge to rule whether 3 M age-bias suit is class action. *Star Tribune,* p. D1. Retrieved from Proquest National Newspaper Premier.

Another threat to economy: Boomers cutting back. (2010, August 15). *Wall Street Journal (Online)*, Proquest National Newspaper Premier.

Audit Bureau of Circulations. (2011). *Circulation averages for the six months ended 9/30/2011: FAS-FAX Report – 9/30/2011*. Retrieved from http://abcas3.accessabc.com/ecirc/newstitlesearchus.asp

Axtman, K. (2010, April 17). Portrait of a long life: He's not retiring – about himself, or from his job. *The Christian Science Monitor*. Retrieved from Proquest National Newspaper Premier.

Beyer, G. (2010, September 12). A career in government. *Wall Street Journal (Online)*. Retrieved from Proquest National Newspaper Premier.

Bonnie, M. R. (2010, December 26). Face lift. Resume boost? *Chicago Tribune*, p. 1.1. Retrieved from Proquest National Newspaper Premier.

Bonnesen, J. L., & Burgess, E. O. (2004). Senior moments: The acceptability of an ageist phrase. *Journal of Aging Studies, 18*(2), 123–142.

Bray, C. (2010, May 21). Heard & scene: Veteran weatherman sues CBS. *Wall Street Journal*, p. A24. Retrieved from Proquest National Newspaper Premier.

Cauchon, D. (2010, December 15). American workforce growing grayer. *USA Today*, p. A1. Retrieved from Proquest National Newspaper Premier.

Channick, R. (2010, November 15). Macy's gets gold star for morale. *Chicago Tribune*, p. 37. Retrieved from Proquest National Newspaper Premier.

Chapman, D. (2010a, March 18). Reinvention beats bitterness. *The Atlanta Journal – Constitution*, p. A10. Retrieved from Proquest National Newspaper Premier.

Chapman, D. (2010b, September 28). 55-older struggle in job search: Older workers face job, life challenges. Numerous reasons why full-time employment is hard to come by. *The Atlanta Journal – Constitution*, p. A1. Retrieved from Proquest National Newspaper Premier.

City/Allegheny County/West/South [derived headline]. (2010, October 30). *Pittsburgh Post – Gazette*, p. B4. Retrieved from Proquest National Newspaper Premier.

Clock ticks on bloated pensions (2011, February 27). Clock ticks on bloated pensions; If legislators don't act to curb pension excesses soon, an outraged public will act. *Merced Sun Star*, p. A6. Retrieved from Proquest National Newspaper Premier.

D'Angelo, P., & Kuypers, J. A. (Eds.). (2010). *Doing news framing analysis empirical and theoretical perspectives*. New York: Routledge.

Dahmen, N. S., & Cozma, R (Eds.). (2006). *Media takes on aging*. International Longevity Center-USA and Aging Services of California. Retrieved from http://www.aging.org/files/public/Media.Takes.book.pdf

Davidson, P. (2010, November 08). 1.2 million people want a job but aren't looking. *USA Today*, p. B1. Retrieved from Proquest National Newspaper Premier.

Denning, L. (2010, September 2). Older U.S. work force has an ugly wrinkle. *Wall Street Journal*, p. C20. Retrieved from Proquest National Newspaper Premier.

Dickinson, A. (2010, April 16). Flag parent about strange behavior. *Chicago Tribune*, p. 2. Retrieved from Proquest National Newspaper Premier.

Dillman, L. (2010, November 13). NBA; inside look into clippers; documents from Baylor's lawsuit tell of his requests of team, and sterling's remarks. *Los Angeles Times*, p. C7. Retrieved from Proquest National Newspaper Premier.

Dolan, M. (2010, May 27). Fired Google manager claims age bias in suit; state supreme court hears arguments in search giant's attempt to have case dismissed. *Los Angeles Times*, p. B4. Retrieved from Proquest National Newspaper Premier.

Dolnick, S. (2011, January 06). New round of worker complaints at Saigon Grill. *New York Times*, p. A22. Retrieved from Proquest National Newspaper Premier.

Dugas, C. (2010, October 25). For many over 55, debt defers dreams. *USA Today*, p. A1. Retrieved from Proquest National Newspaper Premier.

Efrati, A., & Koppel, N. (2010, August 06). Corporate news: Google is set back in age-bias case in ruling with broad implications. *Wall Street Journal*, p. B4. Retrieved from Proquest National Newspaper Premier.

Egelko, B. (2010, July 20). Part of ex-TV reporters' bias suit tossed. *San Francisco Chronicle*, p. C3. Retrieved from Proquest National Newspaper Premier.

Egelko, B. (2011, January 14). KPIX age-bias claim dismissed by judge. *San Francisco Chronicle*, p. C2. Retrieved from Proquest National Newspaper Premier.

Enforce limits on noise from motorcycles. (2011, January 13). *USA Today*, p. A8. Retrieved from Proquest National Newspaper Premier.

Farrell, C. (2010, June 27). Your money. *Star Tribune*, p. D10. Retrieved from Proquest National Newspaper Premier.

Fealy, G., McNamara, M., Treacy, M. P., & Lyons, I. (2011). Constructing ageing and age identities: A case study of newspaper discourses. *Ageing and Society*. doi:10.1017/S0144686X11000092. Retrieved from Cambridge Journals Online.

Felten, E. (2010, June 25). Taste – de gustibus: In Hollywood, new rage against age. *Wall Street Journal*, p. W11. Retrieved from Proquest National Newspaper Premier.

Fernandez, M. (2010, March 30). He opened a door and found a long career. *New York Times*, p. A22. Retrieved from Proquest National Newspaper Premier.

Fishman, T. (2010, October 27). Communities can age gracefully. *USA Today*, p. A11. Retrieved from Proquest National Newspaper Premier.

Funt, P. (2010, August 10). Fuddy-duddy fights back. *Boston Globe*, p. A13. Retrieved from Proquest National Newspaper Premier.

Gallagher, J. (2010, October 17). Workforce changes let employees, firms age gracefully. *Chicago Tribune*, p. 2A1. Retrieved from Proquest National Newspaper Premier.

Garone, E. (2010, May 07). Is age a factor in the teaching field? *Wall Street Journal (Online)*. Retrieved from Proquest National Newspaper Premier.

Gerontology News. (2011). *Journalists promise diverse portrayal of aging in America*. Washington, DC: The Gerontological Society of America.

Giegerich, S. (2010, August 20). Computers, internet daunting for older people; 'it's a jungle,' one job hunter says; CAREERS. *St. Louis post – Dispatch*, p. B1. Retrieved from Proquest National Newspaper Premier.

Global Action on Aging. (2007). *The five year review of MIPAA*. Retrieved from the Internet on December 5, 2011, from http://globalaging.org/agingwatch/events/CSD/MIPAA+5.htm

Goodman, E. (2011, January 1). As they hit 65, boomers redefine aging. *Boston Globe*, p. A11. Retrieved from Proquest National Newspaper Premier.

Grantham, R. (2010a, May 02). Longtime workers go back to school: Career fear sends boomers to college. Three who re-enrolled are optimistic now about their jobs ahead. *The Atlanta Journal – Constitution*, p. D1. Retrieved from Proquest National Newspaper Premier.

Grantham, R. (2010b, November 28). Retirement plans redone: Recession was sudden roadblock for some. Many find career over sooner than expected; others work longer. *The Atlanta Journal – Constitution*, p. D1. Retrieved from Proquest National Newspaper Premier.

Greenhouse, S. (2010, March 04). The job you make. *New York Times*, p. F1. Retrieved from Proquest National Newspaper Premier.

Hailey, C. (2010, October 31). Back to work: Many older Americans are finding 'encore careers'. *Boston Globe*, p. O2. Retrieved from Proquest National Newspaper Premier.

Hsu, T. (2010, October 10). America out of work; 73-year-old not ready for retirement. *Los Angeles Times*, p. B3. Retrieved from Proquest National Newspaper Premier.

Julian, K. (2010, June 30). Where did the summer jobs go? *Pittsburgh Post – Gazette*, p. B7. Retrieved from Proquest National Newspaper Premier.

Kaslow, A., & Tate, P. (2011, February 14). US economy's hidden asset: Older workers. *The Christian Science Monitor.* Retrieved from http://www.csmonitor.com/Commentary/Opinion/2011/0214/US-economy-s-hidden-asset-older-workers

Kever, J. (2010, May 27). HCC police investigated over firing of 3 officers doubts raised about whether the department's chief is qualified CHIEF: EEOC complaints filed by 2. *Houston Chronicle*, p. B1. Retrieved from Proquest National Newspaper Premier.

Klein, E. (2010, November 28). Retirement security on the altar of fiscal responsibility. *The Washington Post*, p. G1. Retrieved from Proquest National Newspaper Premier.

Layton, L. (2010, December 31). Agents sue FBI over its term-limit policy. *The Washington Post*, p. B3. Retrieved from Proquest National Newspaper Premier.

Lee, D. (2010, April 15). LABOR; for many, joblessness is a long-term struggle; A record 44 % of unemployed lost their jobs at least six months ago. *Los Angeles Times*, p. B1. Retrieved from Proquest National Newspaper Premier.

Leland, J. (2010, April 25). For a graying population, care increasingly comes from a graying work force. *New York Times*, p. A14. Retrieved from Proquest National Newspaper Premier.

Lindgren, A. (2010, February 28). If age is a weak spot, don't make it obvious. *The Atlanta Journal – Constitution*, p. G2. Retrieved from Proquest National Newspaper Premier.

Lotan, G. T. (2010, May 16). Transitioning to the do-gooder. *Boston Globe*, p. G6. Retrieved from Proquest National Newspaper Premier.

Meyer, A. (Premier. 2010, November 22). Gray hair looking better to employers. *Chicago Tribune*, p. 19. Retrieved from Proquest National Newspaper.

Miller, V. (2010, May 30). A bigger touch of gray. *Las Vegas Review – Journal*, p. E1. Retrieved from Proquest National Newspaper Premier.

Murphy, E. (2004). *Older people and aging issues in Irish newspapers: A preliminary analysis.* Dublin: Age & Opportunity. Retrieved from http://ageandopportunity.ie/news-publications/working-media

Murray, S. (2010, July 28). U.S. news: Public employees get more benefits. *Wall Street Journal*, p. A2. Retrieved from Proquest National Newspaper Premier.

Nelson, T. E., Oxley, Z., & Clawson, R. A. (1997). Toward a psychology of framing effects. *Political Behavior, 19*(3), 221–246.

Nelson, T. D. (2002). *Ageism: Stereotyping and prejudice against older persons*. Cambridge: MIT Press.

Norris, F. (2010, April 17). In global unemployment, A sea of young faces. *New York Times*, p. B3. Retrieved from Proquest National Newspaper Premier.

Norris, F. (2011, January 15). Older workers are keeping a tighter grip on jobs. *New York Times*, p. B3. Retrieved from Proquest National Newspaper Premier.

O'Connell, V. (2010, July 22). 'Objection!' older lawyers resist forced retirement. *Wall Street Journal*, p. B1. Retrieved from Proquest National Newspaper Premier.

Ohlemacher, S. (2010a, November 19). Raising retirement age would hurt the poor, GAO says. *Pittsburgh Post – Gazette*, p. A5. Retrieved from Proquest National Newspaper Premier.

Ohlemacher, S. (2010b, November 19). Retirement age tweak's downside; lower-income workers at risk, congress is told. *St. Louis Post – Dispatch*, p. C1. Retrieved from Proquest National Newspaper Premier.

O'Reiley, T. (2010, September 30). Republic to pay $3 million. *Las Vegas Review – Journal*, p. D1. Retrieved from Proquest National Newspaper Premier.

Pabst, L. (2010, May 16). Whistleblower. *Star Tribune*, p. B3. Retrieved from Proquest National Newspaper Premier.

Palmore, E. (1999). *Ageism: Negative and positive*. New York: Springer Publishing Company.

Pender, K. (2010, July 11). Jobless boom a boon to lawyers. *San Francisco Chronicle*, p. D1. Retrieved from Proquest National Newspaper Premier.

Pope, E. (2010, March 04). Matching life experience with new careers. *New York Times*, p. F1. Retrieved from Proquest National Newspaper Premier.

Porter, W. (2010, August 19). "Me? retire?" more over-65 s on the job. *Denver Post*, p. A1. Retrieved from Proquest National Newspaper Premier.

Pritscher, J. (2010, May 02). Yes, I'm 26 and yes, I do the hiring. *New York Times*, p. BU9. Retrieved from Proquest National Newspaper Premier.

Pugh, T. (2010, April 02). Job pressure greater on younger workers. *Chicago Tribune*, p. 23. Retrieved from Proquest National Newspaper Premier.

Rampell, C. (2010, May 16). Jobs that are gone for good unemployed workers with obsolete skills struggle as the economy shifts. *Pittsburgh Post – Gazette*, p. A7. Retrieved from Proquest National Newspaper Premier.

Rich, M. (2010, September 26). Older unemployed struggle to find work. *Pittsburgh Post – Gazette*, p. A11. Retrieved from Proquest National Newspaper Premier.
Rudman, D., & Molke, D. (2009). Forever productive: The discursive shaping of later life workers in contemporary Canadian newspapers. *Work, 32*, 377–389. doi:10.3233/WOR- 2009-0850.
Sachdev, A. (2010, February 02). Age bias at law firms an issue again. *Chicago Tribune*, p. 16. Retrieved from Proquest National Newspaper Premier.
Scarborough Research. (2009). *Scarborough writes a refreshing headline for the newspaper industry: Three-quarters of adults are reading newspapers, in print or online*. Retrieved from http://scarborough.com/press_releases/Scarborough-Newspaper-Audience-Readership-November-2009.pdf
Scheufele, B. T., & Scheufele, D. A. (2010). Of spreading activation, applicability and schemas: Conceptual distinctions and their operational implications for measuring frames and framing effects. In P. D'Angelo & J. A. Kuypers (Eds.), *Doing news framing analysis, empirical and theoretical perspectives* (pp. 110–134). New York: Routledge.
Schmich, M. (2010, October 13). PR woman, 72: 'My age is my hook'. *Chicago Tribune*, p. 6. Retrieved from Proquest National Newspaper Premier.
Schultz, E. (2010, April 07). Corporate news: AT&T fights pension suit – potential liability for retirees' plan conversion is estimated at $2.3 billion. *Wall Street Journal*, p. B2. Retrieved from Proquest National Newspaper Premier.
Seltz, J. (2010, September 23). Retirement age, but staying on the job. *Boston Globe*, p. GS1. Retrieved from Proquest National Newspaper Premier.
Semuels, A. (2010a, September 5). America out of work; Mignon Veasley-Fields. *Los Angeles Times*, p. A27. Retrieved from Proquest National Newspaper Premier.
Semuels, A. (2010b, September 19). America out of work/Raul Vargas; fighting a rising sense of futility. *Los Angeles Times*, p. B7. Retrieved from Proquest National Newspaper Premier.
Seniors don't just take, but they also give back. (2010, December 20). *USA Today*, p. A8. Retrieved from Proquest National Newspaper Premier.
Sestak, J. (2010, September 23). Preserve social security my opponent would privatize it, although he denies it now. *Pittsburgh Post – Gazette*, p. B7. Retrieved from Proquest National Newspaper Premier.
Shapira, I. (2010, April 03). Working hard, or hardly working?; study casts doubt on millennials' effort, but they say it's not all about a 9-to-5 life. *The Washington Post*, p. B1. Retrieved from Proquest National Newspaper Premier.
Simmons, A. (2010, August 16). Amid cuts, sheriffs value more senior volunteers: Fulton, Gwinnett look to start retiree programs as others in area grow. *The Atlanta Journal – Constitution*, p. B1. Retrieved from Proquest National Newspaper Premier.
Sixel, L. (2011, January 14). Adult club waitress settles age bias case woman, then 56, sued over firing, saying bosses called her old, forgetful. *Houston Chronicle*, p. D3. Retrieved from Proquest National Newspaper Premier.
Smiley, E. (2011, January 03). Don't hire based on looks. *Pittsburgh Post – Gazette*, p. A8. Retrieved from Proquest National Newspaper Premier.
Stafford, D. (2010, November 14). Keep an open mind when the boss is younger. *Pittsburgh Post – Gazette*, p. F1. Retrieved from Proquest National Newspaper Premier.
Staples, G. B. (2010, February 2). Volunteers love their jobs: Despite neighborhood changes, grandmothers stay put, inspire kids. *The Atlanta Journal – Constitution*, p. B1. Retrieved from Proquest National Newspaper Premier.
Steffy, L. (2010, October 24). A senior webmaster is keeping current steffy: Older entrepreneurs are a new wave. *Houston Chronicle*, p. D1. Retrieved from Proquest National Newspaper Premier.
Sulzberger, A. (2010, December 22). Helping hallmark greet season after season. *New York Times*, p. A18. Retrieved from Proquest National Newspaper Premier.
Trumbull, M. (2010, March 31). Obamas get to work from home. so should you, they say. *The Christian Science Monitor*. Retrieved from Proquest National Newspaper Premier.

Turner, A. (2011, January 28). TO THRIVE PAST 65 GeezerFest celebrates workers who keep going strong at advanced ages FEST: Retirement can wait. *Houston Chronicle*, p. B1. Retrieved from Proquest National Newspaper Premier.

United Nations. (2003). *Political declaration and Madrid international plan of action on ageing, Second world assembly on ageing, Madrid, Spain, 2002*. New York: United Nations Department of Public Information.

Urban, C. D. (1999). *Examining our credibility: Perspectives of the public and the press*. Retrieved from http://www.media-diversity.org/en/index.php?option=com_content&view=article&id=340

Vanderkam, L. (2011, January 5). This isn't grandpa's retirement. *USA Today*, p. A7. Retrieved from Proquest National Newspaper Premier.

Wakin, D. (2010, July 14). Jury selection continues in music critic's suit. *New York Times*, p. C3. Retrieved from Proquest National Newspaper Premier.

Ward, B. (2010, November 13). Not the retiring kind. *Star Tribune*, p. E1. Retrieved from Proquest National Newspaper Premier.

What's the big idea? (2011, January 16). *The Washington Post*, p. B2. Retrieved from Proquest National Newspaper Premier.

Wolf, R. (2011, January 28). When retiring means giving back. *USA Today*, p. A1. Retrieved from Proquest National Newspaper Premier.

Chapter 5
The Evolution of Retirement as Systematic Ageism

Lynn McDonald

Introduction

Researcher Alan Walker stated that "…retirement is both the leading form of age discrimination and the driving force behind the wider development of ageism in modern societies…retirement maybe seen as an age discriminatory social process designed to exclude older people *en masse* from the workforce"(Walker 1990, p. 59). A number of scholars echo his sentiments (Bytheway 1995; Friedman 2003; Longino 2005; Palmore 2005; Townsend 2006), while others see retirement as 'justified' ageism or at least question whether retirement is a form of ageism that gave birth to general ageist ideas in developed countries (Grattan 2002; Macnicol 2006, 2010). In this chapter it is argued that retirement is an institutional form of ageism – both negative and positive – depending on the era in question, beginning with the inception of retirement in the twentieth century through to retirement today. The approach taken uses the political economy framework that retirement is driven by changes in the economy and supported by the enlightened self-interest of various political and economic groups (McDonald and Wanner 1990). The chapter will draw on the North American experience, specifically Canada, to illustrate the role of the economy in determining the changing meaning of retirement as it responds to the requirements of the market, supported by stereotypes of older adults that are attached to the various market changes over time. Unlike other accounts of retirement and ageism, this account accepts that the evolution of capitalist economies has both *incorporated and marginalized* older workers over time via pension and retirement polices (Binstock 2010), but does not accept economic orthodoxy "that the invisible hand of

L. McDonald, Ph.D. (✉)
Faculty of Social Work, Institute for Life Course and Aging, University of Toronto,
222 College Street, Suite 106, Toronto, ON M5T3J1, Canada
e-mail: Lynn.mcdonald@utoronto.ca

P. Brownell and J.J. Kelly (eds.), *Ageism and Mistreatment of Older Workers: Current Reality, Future Solutions*, DOI 10.1007/978-94-007-5521-5_5,

the market transmutes individual acts of selfishness into socially desirable collective outcomes" (Cassidy 2009, p. 8). The history of retirement, considered here as the handmaiden of the ever-changing economy (Halliwell 2009; Phillipson and Smith 2005; Walker 2006), is littered with examples of undesirable outcomes for older adults based on systematic ageism.

Setting the Parameters

Retirement, as a social institution, emerged in modern industrialized societies at the beginning of the twentieth century. Scholarship into retirement paralleled its development and began in earnest in the late 1940s and early 1950s. However, because of the rapid evolution of retirement (McDonald 2011), there is some uncertainty about the relevance of research conducted in earlier times to retirement today (McDonald and Wanner 1990). Just as social scientists think they have explained the transition from work to retirement and its causes and consequences, the concept shifts according to different economic conditions, policies and individual preferences that require revised explanations. For example, while retirement may have been initially developed as a reward for loyal older workers to encourage life-long attachment to employers, in a recession, early retirement may be used as a mechanism to discard older workers from the labour force as quickly as possible. These variations on retirement reflect the idea that retirement as an institution was designed to move the older worker out of the labour force in a systematic manner without causing undue financial upset, while solving the societal problem of what to do with an aging labour force (Chappell et al. 2008). In other words, to a certain extent, the nature of retirement depends on the size, composition and requirements of the labour market at any given time (McDonald and Wanner 1990).

In light of this typical definition of retirement, it is difficult to ignore the social construction of retirement as an ageist institution, albeit a useful one to some sectors of society. The progenitor of this definition, Robert Atchley (1980, p. 264), reinterprets his own institutional view of retirement 13 years later in a more critical light to show how removing workers from the labour force serves the interest of capitalists and is of questionable value to society. He argues that retirement, as an institution, ensures that society has no obligation to provide jobs for older workers who may wish or need to be gainfully employed. In addition, he claims that the institution of retirement reinforces the view that older workers are less productive then younger workers (Atchley 1993). His more recent views represent those of most anti-ageists (Macnicol 2006).

Ageism here refers to institutional ageism, the "missions, rules and practices that discriminate against individuals and/or groups because of their older age" (International Longevity Center 2006, p. 21). It also includes both positive and negative ageism as first developed by Palmore (1990) and later Cole (1992) who labeled both forms the polarization of ageism. Notably, Erdman B. Palmore defines ageism as "any prejudice or discrimination against or in favor of an age group"

(1990, p. 4). In a study of polarized ageism, representations of older adults in a national newspaper in Canada (Rozanova et al. 2006) were those of 'successful' aging often seen as a positive view – "a lifelong process optimizing opportunities for improving and preserving … independence; quality of life; and enhancing successful life-course transitions" (Peel et al. 2005, p. 298). Although optimistic, assumptions based on positive ageism may lead to responses that are just as biased as negative ageism and can do as much harm (Binstock 2010; Townsend 2006). Thus, those who don't 'age successfully' are likely aging failures, notwithstanding their genetic make-up, accidents and poor health (Rozanova et al. 2006).

At the same time, the ageism that retirement embodies is often implicit, meaning that thoughts, feelings and behaviours toward older adults exist without conscious awareness, with the assumption that the ageism is the basis of most interactions with older individuals (Levy 2001, p. 578). This would include the realm of 'justified' ageism where, the attributions of discrimination are minimized by the account-giver to fit with their own images of fairness (Levy 2001; McVittie et al. 2003). Many account-givers offer rationalizations for retirement that have little to do with the facts or the older adults they represent. For example, some scholars suggest that older workers are found in declining labour market sectors and therefore have few alternatives but to retire, an idea that appears normal enough at first glance. The problem, however, is that this observation is age based and is a stereotype that implies all older workers cannot be educated or develop new skills, they are all incapable of moving to other industries and retirement is just another name for unemployment (Finnie and Gray 2011; Pignal et al. 2010; Townsend 1981). Finnie and Gray investigate the sources of income and the labour market choices of long-tenured older workers in the 5 years following layoff in Canada, and find that 66% of those aged 60–64, choose to, or are forced to retire due to poor re-employment prospects (2011, p. 19), actually supporting retirement stereotypes.

Since retirement is addressed at the institutional level, the theories of 'acquiescent functionalism' of the early gerontologists expressed in certain micro theories like disengagement, continuity and activity theories are inconvenient as they ignore structural forces that affect retirement (Cumming 1963; Donahue and Tibbitts 1957; Parsons 1942). The same can be said for the human capital arguments that rely on individualistic explanations of retirement based on the acquisition of knowledge and skills (McDonald and Wanner 1990). The concept of 'acquiescent functionalism' used by Townsend refers to a body of thought, "that attributed the causes of the problems of old age to the natural consequences of physical decrescence and mental inflexibility, or to individual failures of adjustment to ageing and retirement, instead of the exertions of state economic and social policy partly to serve and partly to moderate market forces" (Townsend 2006, p. 164). Although more a perspective then a theory, political economy suggests that aging is a socially constructed process that is conditioned by one's location in the social structure and the economic, political and social factors that affect it (Estes et al. 1984; Phillipson 1982).

The policies that demarcate retirement generally include retirement and pension policies. Retirement policy is primarily about the age of eligibility and other requirements that regulate when and how people retire. Pension policy is what makes retirement possible: how the accumulation of pension benefits is designed and financed. "The two policies are strange bedfellows even though a change in one policy noticeably affects the other. In Canada, retirement policy is reformed largely through the back door of pension policy that tends to be sensitive to market pressures, mainly to remedy whatever the country's current fiscal imperative is" (McDonald and Donahue 2011, p. 409). Both sets of these polices usually contain stereotypes of the aged that provide 'legitimate' reasons for the ageist aspects of retirement.

This perspective of retirement is not common, particularly in North America where questioning the functioning of market economies is not the norm. The markets are seen to be non–ageist because they are considered to be more efficient than any other system, they are self-correcting and "scientific" (e.g. general equilibrium theory), suggesting that they are value-free (Cassidy 2009). A close read of many of the justifications for retirement over time are based on the ebb and flow of the market according to the demand for older workers, a process free of systematic ageism because of the 'natural' operation of the markets in a capitalist economy.

Age Discrimination in Retirement[1]

There are as many histories of retirement as there are developed countries with retirement programs and pensions. All the histories have, however, several commonalities. Retirement is a social construction that took root in modern industrialized societies near the beginning of the twentieth century and became a well-established social institution in western societies following World War II (Achenbaum 1978; Atchley 1982; Fischer 1978; Graebner 1980; Haber 1978; Haber and Gratton 1994; Phillipson 1999; Snell 1996). Secondly, retirement has emerged as the last segment of the life course and helps to define 'old age' as a distinct chronological phase apart from education at the beginning of the life course and the middle phase, which is devoted to work. Retirement, then, is not only the principle gateway to later life but is the conduit that links the institutional structures of work and non-work.

Two historical models are used to explain how retirement came to be, the "…impoverishment model, evident in the history of welfare…" (Haber and Gratton 1994, p. 88), where the catalyst causing mass retirement of the old was industrialization and the competing model, the "social security model" developed in the United States where the establishment of social security is argued to have led to mass retirement (Haber and Gratton 1994). A version of these two models appears in Macnicol's (1998, 2008) work in his supply and demand accounts of retirement, and their combination, where he favours the decline in the demand for older workers in advanced industrial economies, and provides evidence for this process in Britain that started in the 1890s. Doubtless, the history of retirement will be context specific depending

[1] In reviewing men and women's behavior patterns from pre-industrial times until the 1960s, women were invisible in the development of retirement because its evolution into a social institution occurred within the context of the labour market where women were least likely to be found (see McDonald 2002; 2006 for a history of women's retirement in Canada).

on the country in question and most likely is more nuanced beyond an amalgamation of both views. Currently, North America has a propensity for favouring individual behavioural explanations and the roles of policies in corporations (Atchley 1976; Binstock 2010; Hardy et al. 1996; Marshall and Marshall 2003; McDonald and Wanner 1990; Quinn et al. 1990), while Europe leans toward the function of public policies in creating and sustaining retirement (European Commission 2007; Kohli and Rein 1991; OECD 1995; Phillipson 2004; Platman and Taylor 2004).

Agrarian Times

Retirement did not exist in most countries prior to industrialization, and most people adapted their work to their changing capacities by working until they were no longer capable. Historian Andrejs Pakans (1989), argues that people stepped down from their work in traditional Europe but the process of withdrawal was informal and could be long or short depending on the person's health and/or wealth. In agrarian Canada, a farmer could slowly diminish his workload by delegating the more demanding tasks to his sons or sons-in-law and still maintain control of the process because he owned the land that the children wished to inherit. Because the history of retirement is primarily one of men's retirement, little is known of women although they probably stepped down in tandem with their husbands. If the male worker was a non-property owner in the employ of others, the tendency was to continue work until physically impossible. At the time, the relations of labour were characterized as paternalistic, a more personal type of relationship between employer and employee that preceded the more impersonal relationships of industrial capitalism (Pentland 1981, p. 26). According to historical accounts of the nineteenth century workplace, older workers were re-assigned to jobs that suited their abilities so they could work as long as possible (Macnicol 1998; McDonald 2002; Snell 1996). In short, biological aging provided the impetus for the gradual withdrawal from work for most in a pre-industrialized Canada.

Pensions were unheard of for the majority of older people but did exist for some specific groups in the private sphere. In Canada in the eighteenth-century, the Hudson Bay Company did make some provision for "...the servants in case of sickness and old age" and allowed a pension in "any case that was deserving" (Simpson 1975, p. 182–7). Older soldiers in far-flung outposts like those in the British garrison in Canada were paid pensions, which served to attach them to Canada while the younger soldiers deserted to the United States. Historians Morton and McCallum note that, "Such men would endure the privations and discipline of military life in return for a secure old age" (Morton and McCallum 1988, p. 6).

The Grand Trunk Railway established one of the first industrial pension plans in Canada in 1874.[2] Pre-occupied with the job of imposing discipline and stability on

[2] The Superannuation Act of 1870 provided an early occupational plan for federal government employees. The purpose of the plan was to "... get rid of persons who had arrived at a time of life when they could no longer perform their work efficiently" (Morton and McCallum 1988, p. 6).

armies of employees in a far-flung enterprise like a railway, the managers turned to military solutions such as uniforms, hierarchies, strict regulations and pensions with compulsory retirement at age 65 to reinforce the worker's commitment to the company (Morton and McCallum 1988). Pensions at this time were a gift to insure that ageing male workers did not jeopardize their relations with their employer (e.g. going on strike), in return for a secure old age.[3] A Royal Commission, struck in 1919 to deal with the worst strikes in Canadians history, was told by BESCO, a Cape Breton corporation, that a pension was a *management tool* to encourage the loyalty of employees to reduce labour turnover, and make it easier to rid the company of older workers (Morton and McCallum 1988, p. 11). An Imperial Oil employee was more blunt, stating that corporate welfare (a pension), "is not philanthropy and it is not benevolence: it is a cold-blooded business proposition" (Morton and McCallum 1988, p. 11).

Although the railway executives stressed their concern for the welfare of the older worker, their motives were not especially benevolent (Haber 1978, p. 81). At this time the seeds were sown for the relationship between old age and incapacitation, the idea of an age-related retirement and the idea of a reward for the 'deserving old,' – all ageist ideas borrowed from the military that operated to serve management in their attempt to control a difficult labour market via corporate policies (National Union 2007).

In the 1880s, Bismarck, also relying on a military metaphor, added fuel to the fire when he called for state pensions for working class men in a rapidly industrializing Germany:

> The State must take the matter into its own hands, not as alms giving but as the right that men have to be taken care of when, from no fault of their own, they have become unfit for work. Why should regular soldiers and officers have old age pensions, and not the soldier of labour? (Quoted in Donahue et al. 1960, p. 351).

Canadian response to Bismarck's proposition was not one that flagged his ageist ideas about inevitably unfit soldiers of labour but about the dangers of social insurance that could wreak havoc in the labour force. Professor Mavor of the University of Toronto, representing the voice of many, saw all pension schemes as a method to subsidize low wages and objected to social insurance on the grounds that malingering would make the system unworkable (Wallace 1950).

Industrialization

With the acceleration of industrialization a number of factors converged that contributed to the growth of retirement that promoted a negative ageist view of older adults. "Scientific management" (Taylor 1947) emphasized the division of the

---

[3] In 1910 in the course of a bitter strike, the Grand Trunk Railway wiped out the pension rights of the workers who had struck the company, only to return them in 1923 . when the grand Truck became part of the government-owned Canadian National Railways.

whole production process into small, quickly learned and routinized operations with quantity, quality and time standards for each task. At the time, this trend reduced the need for skilled workers, a significant characteristic of older workers. As a result, the older worker lost considerable prestige because the skills associated with his craft were no longer valued. At the same time, the wear-and-tear theory of aging gained ascendency (Achenbaum 1978), supporting the principles of scientific management. Although a number of business leaders and physicians had noted the uselessness of older people earlier, the renowned Dr. William Osler, (originally from Ontario), stated in 1905 that men above 40 years of age were comparatively useless and secondly, that men above 60 years of age were useless and there would be "…incalculable benefit … in commercial, political and professional life, if as a matter of course, men stopped work at this age" (The Globe 1905, p. 1). For good measure he recommended "peaceful departure by chloroform" 1 year after retirement (The Globe 1905, p. 1). This appeared in his farewell address at John Hopkins University on his way to the Medical School at Oxford.

The wear-and-tear views were further bolstered by the adoption of machine technology and the speed at which it operated. At the time, the unions were agitating for a 9 hour day and higher wages while management wanted a return on their investment in expensive machinery. A suitable compromise between the two was to increase the work rate, to the supposed detriment of 'useless older workers' that gave workers shorter hours and management more profit. Alongside the growing belief that older workers were useless was the added observation by economists that the numbers of older people were growing. In 1871, those aged 45–64 represented 11.1% of the total population, but by 1931 they made up 16.8% (Bryden 1974, p. 30). An increasingly common view was that older workers and their lack of productivity were a national menace to overall economic productivity as stated frankly in the Canadian Senate (Parliament of Canada 1926–27).

Because of the all too effective lobbying of social reformers, the majority of older persons came to be associated with poverty as a result of being forced out of the labour force (Snell 1996). Comments at the time, such as that of the Trades and Labour Council of Nova Scotia, bemoaned that workers were assigned "to the 'scrap pile' at the first sign of age" (Quoted in Snell 1996, p. 28). Many organized groups, like the Social Service Congress and the Trades and Labour Congress of Canada, demanded old age pensions for poor, older people. The witnesses to the various parliamentary committees set up between 1911 and 1924 provided similar evidence as to the ravages of industrialism and the inability of poor families to provide for their aged members (House of Commons Debates 1921, p. 3860). In 1927 the Old Age Pensions Act (OAP) was reluctantly passed at the urging of J.S. Woodsworth and A.A. Heaps, two Labour members of the House of Commons who had promised support for a minority government of the day. The Act was to supplement, not replace, the income of workers and was set at subsistence levels to encourage thrift (McDonald and Wanner 1990).

The pension restricted to those 70 years of age and over did not encourage the retirement of men or the few working women, given that it was a piece of "social assistance" legislation conceived as a classic form of poor relief for those already forced out of the labour force. The overly stringent administration of the Act and the miserly amounts awarded meant it could never provide income security in old age. The systematic stereotypes of older adults enshrined in the Act was that of a group of, useless, profligate older adults, jettisoned from the labour market and living in abject poverty.

A significant postscript to these stereotypes is found in the invention of mandatory retirement. Management practices from the United States filtered into Canada and were rapidly adopted in the 1920s. In America, the innovation of the Ford Motor Company of rewarding seniority through annual bonuses and the introduction of the wage ladder became particularly attractive during labour strikes in the early 1900s. By creating a complex set of graduations among indistinguishable jobs within the factory, Ford provided semiskilled workers with an artificial hierarchy to climb, one that was tied to seniority rather than skill. With the creation of the internal labour market the seniority schemes necessitated a cut-off point. Older workers became inefficient because they received considerably more money than younger workers for doing the same work and, as a result, an important reason for mandatory retirement was invented (Haber and Gratton 1994, p. 109; McDonald 2011).

Mandatory retirement, which stipulates an age-mandated end to employment, has elicited considerable condemnation as an exceptional example of age discrimination that has been well documented in many quarters (Gillin and Klassen 1995, 2005; Ibbott et al. 2006; Klassen and Gillin 1999; Lowe 1991). The attacks on mandatory retirement are all based on some form of negative bias against older adults. The civil rights side of the debate focuses on the comparison between the social benefits of a right to contract with the social benefits of a right to be free from age discrimination. Others stir up the intergenerational debate that compares the interests of older and younger workers; still others are concerned with evaluating the productivity of older workers compared to the relative efficient mechanism of shedding all older workers from the labour force through mandatory retirement.

It is noteworthy that historically mandatory retirement was not a policy at any level of government in Canada but rather was developed in the negotiated collective agreements and formal personnel procedures of many Canadian businesses and industries (Ibbott et al. 2006). In short, it was developed as a tool to securely attach workers to the firm in times of strife by 'buying their loyalty' but also to get rid of them when they were older, usually at age 65, and too expensive to maintain. Business developed retirement through the provision of industrial pensions for reasons other than goodwill and then had to adjust the somewhat flawed policy via the introduction of mandatory retirement to prevent the runaway costs incurred by older workers. On one level, it is ingenious to reward someone simply because they aged 1 year. However, to be caught in an ageist, self-created trap and solve the problem by arbitrarily removing older workers from the labour force based on age, bespeaks compounded ageism at its best.

Haber and Gratton argue that the internal markets and seniority systems truncated by mandatory retirement were not instances of conscious ageism, but rather management solutions to labour problems (Haber and Gratton 1994, pp. 106–107). Mandatory retirement was indeed a management solution as has been argued above but it was also an *ageist* management solution, possibly unconscious, but harmful to many. During the Great Depression, with a quarter of the workforce unemployed and at least 15% of the nation on relief, many older workers found themselves in difficult circumstances. They faced growing mandatory retirement provisions in firms, a lack of jobs and no private pension coverage (McDonald 2002). A study by the National Commission on Employment in 1937 found that fewer than 8% of Canadian businesses actually had a pension plan, and a survey by Queen's University in 1938 estimated that 70% of Canadians had no old age income protection whatsoever (Morton and McCallum 1988).

The Consolidation of Retirement in Canada

The negative ageism experienced by retirees took a significant turn in favour of positive ageism – not with the introduction of social security in 1951 – but well into the 1960s when the pension system was consolidated. Unlike other Western nations such as the USA, a universal government pension was not introduced into Canada until the economic expansion following World War II. Not surprisingly, there were a myriad of complex forces that influenced the formulation of the new pension act of 1951 but the reasons were mainly economic. The common threads that run through most explanations of universal retirement included: first and foremost, the wish of Canadians to avoid the economic disasters of the Depression in its period of reconstruction following WWII; the rise of Keynesian economics that provided a blueprint to reach this goal via government responsibility for economic growth and social security and the human agency of labour, social activists and senior citizen groups that fought for pension changes stemming from the flaws of the 1927 pension plan. The Old Age Security Act, financed and administered by the federal government, paid a pension of $40 per month to all Canadians at the age of 70, with no means test involved. The belief at the time was the economic growth following the war would generate enough financial resources to underwrite the costs of the pension (McDonald and Wanner 1990).

'Compassionate ageism' (Binstock 2010) did not appear in Canada in 1951 like in the USA after the establishment of Social Security in 1935. Even with the passage of the Social Security Act of 1951, the perceived association among uselessness, poverty and retirement lingered on from the beginnings of industrialization, as did mandatory retirement. The underlying ideology of the Act was still that the pension was a measure against destitution in old age as a result of retirement. If public opinion polls, popular media accounts and 'expert advice' are an indication of the times, in the public mind retirement was well entrenched and decidedly ageist. The popular view at the time showed that retirement was considered a chronological guillotine that assured a drop in income, the onset of boredom and a threat to health, if not death (Financial Post 1957, July 6; Labour Gazette 1957; Macleans 1961).

It wasn't until the Old Age Security Act was restructured through the 1960s into the 1970s and the introduction of the wage-based Canada/Quebec Pension Plan in 1965 that retirement was secured as a significant social institution that was considered a 'normal' part of the life course.[4] These developments happened in a stable economy that could afford social security in the form of universal pensions and health care. The government took over primary responsibility for retirement income, the concept of retirement was separated from the concept of poverty, public pension benefits became 'deferred wages' to which people were entitled due to their contributions, and withdrawal from economic activity took place in advance of physical decline.

Within this environment in the mid 1960s benign ageism or the 'compassionate ageism' noted by Binstock (1983) flourished. The changes noted above represented an about-face in ideology from an emphasis on the individual responsibility of the older adult for his or her own retirement to an emphasis on the deserving older adult who should be collectively insured against the financial risks associated with a fixed-age retirement. The benign ageism embraced stereotypes of the older person as worthy of government support at age 65, but still frail and dependent, a double-edged sword. On the one hand, older workers benefited from the myriad of programs devoted to them, like provincial pension top-ups and the Spouses Allowance, not to mention the good-will generated in the Canadian public by the positive ageist stereotypes (McDonald 1995). On the other hand, the economic dependency of older adults was substantially increased (Walker 2000), according to what Townsend identified as 'structured dependency' the hallmark of the political economy approach. Townsend argues that during the final decades of the 20th century, older people were treated, according to the research, as more dependent than they really were or needed to be and that the emerging institution of retirement (among others) contributed to this dependency (2006, p. 161). As of 1984, the Canadian government provided 51.3% of individuals' income over age 65 (National Council of Welfare 1984, p. 42) and the number was still close to 50% at the beginning of the twenty-first century. This dependency was as widespread in the USA (Binstock 2010) and the UK (Walker 2000). Part of this dependence includes the belief that the income needs of older retired people are lower than for those in the labour market, an age stereotype that justifies setting public pension rates below or at poverty levels in many countries including Canada (Butricia et al. 2009; Walker 2000). While Canada frequently advertises that the income-in-retirement-problem has been solved, it is somewhat astounding that the poverty level for unattached people over age 65 dropped from a massive 69.1 to a disappointing 57.7% between 1969 and 1982.

[4] The pension system of Canada has three pillars: the first pillar consists of public plans (Old Age Security, the Guaranteed Income Supplement, and the Canada/Quebec Pension Plans for paid workers); the second, employer-sponsored plans (RPPs, deferred profit-sharing plans and group registered retirement savings plans [group RRSPs]); and the third, personal savings – including registered retirement savings plans (RRSPs) (Baldwin 2009; Gougeon 2009).

Conservatism and a 'New' Retirement

Early Retirement and Masked Ageism

It did not take long for a change in the format of retirement, again in accordance with the demands of the labour market and global economic pressures. With the oil crises in 1973 and 1979, and the two recessions in 1982 and 1990 that hit Canada and other developed nations, many employers, both in the private and public sectors began to downsize their work forces. This reduction is frequently implemented through offering early retirement packages or incentives to older workers (among other approaches) (Foot and Gibson 1993; McDonald et al. 2000; Siroonian 1993). In Canada in the late 1980s and early 1990s approximately 39% of men aged 55–64 and 12% of men over age 65 retired because of early retirement provisions. In the United States during the 1980s, 40% of American firms with more than 1,000 employees reduced their work forces through early retirement incentives (Meier 1986).

There are three ageist issues intrinsic to early retirement when it becomes the standard for retirement. First, the growth of early retirement redefined the length of old age by beginning at a lower age. The average retirement age in Canada of age 62 in 2009 occurs well before full pensions are activated at age 65 (McDonald and Donahue 2011). Canadians spend, on average, 21.9 years in retirement compared to the French who are expected to spend 25.8 years and the Americans who spend 19.4 years in retirement, a fairly lengthy time compared to retirement when life expectancy was only 68 years in the 1950s (OECD 2010). The lower age-entry into retirement has served, in turn, to reinforce the devaluation of older people in the labour market. The research on discrimination against workers indicates that when workers are labeled older, they become potential targets for prejudice and discrimination related to ageing (Duncan 2001; McCann and Giles 2002) so that discrimination starts at earlier ages. A recent study in Belgium shows that self-categorization as an "older worker" is related to negative attitudes towards work (stronger desire to retire early, stronger inclination towards intergenerational competition) while the perception that the organization does not use age as a criterion for distinguishing between workers, supports positive attitudes towards work (e.g. higher value placed on work) (Desmette and Gaillard 2008). This ageism in the organization is all the more glaring because the workers are actually younger, healthier and more educated than the older workers were in earlier times when 70 was the accepted age for retirement.

Secondly, targeting older workers as a group for early retirement is certainly systematic ageism, possibly of the positive kind since many want to retire early – 47% of men aged 45–59 years in Canada in 2002 (Schellenberg 2004). It is important to note, however, that such plans are offered primarily to workers who tend to be higher paid, better educated and hold professional and managerial positions (McDonald et al. 2000; Schellenberg 1994), challenging the stereotype that all workers received these packages. Using early retirement to ' manage' older workers out of the labour force helped lay the foundation for the belief that all older workers were well-off since they could afford to retire early and the rates of early retirement were so high older adults had the potential to become a burden on the social security system in the future (OECD 1988; Laczko et al. 1988).

Thirdly, one of the newer forms of early retirement is actually involuntary retirement, which is often masked behind more socially justifiable reasons such as "intended" early retirement plans (Rowe and Nguyen 2003; Schellenberg et al. 2005; Tompa 1999; Wang and Shultz 2010). This implicit form of ageism has been well hidden although there is a growing consensus that some older workers have accepted early retirements under duress by colleagues, unions and society at large to "stand aside and make way for younger people, or because they feared for their longer-term security in their jobs and estimated that the benefits offered now were superior to those that might be available when facing redundancy later" (OECD 1995). Using the General Social Survey in Canada in 2002, Schellenberg and colleagues (2005) found that 30% of early retirees reported they were *involuntarily* retired. In another study of unexpected early retirement using a national survey of workers not in the labour force, it was found that an early retirement package did not have a significant impact on retirement income. With the advent of early retirement, Canadians seemed to have created a 'retirement gap' where those forced to retire earlier than planned were out of "sync" with the social safety net because they were not age eligible (McDonald et al. 2000).

"Job Snatchers" and "Greedy Geezers"

The convergence of a number of factors – weak economies, the aging population and the rise of the political right – have been identified by many authors as the impetus behind the assault on the welfare state in western nations which began in the early 1990s in Canada (Midgley 1997; Teeple 1995; Workman 1996). As the preceding brief history of retirement in Canada has shown, the manipulation of factors that influence the labour force behaviour of older workers via pension and retirement polices has always been considered a legitimate means for managing the size and composition of the labour force and for thinning the ranks of corporations, even though here, the manipulation is viewed as systematic ageism. It seems that in the next evolution of retirement, its management was also a legitimate means for reducing the national deficit, the pressing economic crisis of the 1990s.

There was a growing voice issuing from privileged Canadians that older people should contribute more to reducing the deficit (Globe and Mail 1995). In a study that surveyed Canadians on its attitudes toward government, its priorities and its spending, it was found that better-off Canadians were far more likely than others to think that their senior citizens should do more to contribute to fighting the deficit (Ekos Research Associates 1995). This perspective was fueled by claims from government and business that poverty among older people had all but been eradicated and, since retirees had enough money, Canadians could expect them to sacrifice in the name of the deficit. For example, the Canadian Institute of Actuaries, using the most miserly poverty measure available, decided that "... poverty among the aged has been eliminated " (Canadian Institute of Actuaries 1995, p. 17). This message was a politer version of the "greedy geezer" argument (Binstock 1994) promulgated in the United States, which made it easier to politically market slashes to Old Age Security, the Guaranteed Income Supplement, or the Canada/Quebec Pension Plan.

The apocalyptic demography arguments (Robertson 1991) also still held sway. There were too many older workers, taking way too many jobs from too many unemployed younger persons (the baby boomers). Older workers were clogging the higher echelons of company hierarchies and they were attenuating the career mobility of those who followed on their heels. These arguments represented a softer version of the "job snatcher" discourse found in France (Guillemard 1983) and the notion that older workers must make way for younger workers. The demographic load (the ratio of pension beneficiaries to workers) is one of the apocalyptic scenarios that strongly pointed to changes in contributions to, and reductions in, the C/QPP and Old Age Security. Add the noticeable "graying" of the Canadian budget (over 20 billion dollars were transferred to older adults in pensions at this time), and it became almost routine to ask if Canadians had the right to impose such a large economic obligation on future generations – the milder version of intergenerational conflict discourse found in the United States. Superimpose on this scenario the 1991 decision of the Supreme Court of Canada upholding mandatory retirement in those provinces with no human rights legislation protecting older workers, and Canada had unashamedly sanctioned age discrimination (Chappell et al. 2008). Once again, pension policy was available for change to bolster the economy and uncomplimentary and revised stereotypes of older people as prosperous and selfish burdens on the system supported the attack on social security. As Robert Binstock argued in the 1990s, the long-standing compassionate stereotypes of older persons were undergoing a substantial reversal (1994, p. 727) and a new depiction of older people as "ruining the nation" was solidified is his latest assessment (Binstock 2010, p. 576).

The reinstated negative stereotypes of older persons and their retirement justified the retrenchment of the welfare state that was subject to an ambitious effort to scale back Old Age Security benefits in the mid-1990s, an effort ultimately abandoned (Battle 1997; Beland and Myles 2008; Little 2008; McDonald 1997; Weaver 2004). The only feasible possibility for change at the time was the Canada Pension Plan (Little 2008), which underwent fairly significant adjustments such as increased contribution rates, de-indexation of the pay-roll tax exemption level, the reduction of disability benefits, and the creation of an investment board for a hybrid pay-as-you-go scheme (Beland and Myles 2005; Hicks et al. 2005).

The Latest Version of Retirement

> Canada's aging work force hasn't saved enough to retire. Pension benefits are being slashed, employees are working longer, the elderly are selling their homes and going back to work. (Globe and Mail 2009)

The observation from the *Globe and Mail* in 2009 following a world economic downturn is the latest retirement crisis identified by the economic sector, the public, and the media. The 2008 meltdown of financial markets resulted in the recent loss of over $200 billion of household wealth in Canada (Abbott et al. 2009). The trickle of companies retreating from their pension obligations accelerated with the global recession and led some companies into bankruptcies (e.g., Nortel, January 2009). In the process, the recession exposed the underfunding of many pension plans (e.g., Air Canada), with estimates suggesting a $50 billion deficit in corporate pension funds (Globe and Mail 2009). Although the absolute number of members in employment pension plans continued to grow from the addition of women (Moussaly 2010; The Daily 2010), in relation to the size of the employed workforce, membership has been declining since the late 1970s and has fallen from 46.1% of paid workers in 1977 to 38.3% in 2007 (Baldwin 2009; Mintz 2009; The Daily 2010). However, private pension plans grew from 23% in 1992 to 32% in 2006 as a proportion of the average retirement income of Canadians 65 and older (Gougeon 2009), underscoring their growing contribution to retirement income.

Although the trend to later retirement had started prior to the economic downturn, the recession added some urgency to ensuring the new retirement was a later retirement, aided and abetted by the negative stereotypes noted above. The financial industry did add a few more stereotypes to their armamentarium in trying to stabilize the economy, the first being that it was the fault of the older person that they had not saved for their retirement (mainly because of financial illiteracy), the second, that the older person must assume more individual risk for their own retirement income because they were profligate, and last, the unchecked and selfish stampede to early retirement (methodically used in labour relations) had to be reversed.

In the first instance, it is somewhat surprising to lay many of the problems of the "Great Recession" (Rix 2011) at the feet of older adults who experienced widespread financial losses in their life savings, investments and private pension plans that were totally destroyed through no fault of their own (Abbott et al. 2009). For certain, those nearer retirement are more likely than the already retired to be negatively affected because they are still in a labour market distinguished by a "jobless recovery", and they have little time for an economic resurgence in order to regain lost ground in their private pensions and investments (OECD 2009; Sass et al. 2010). The "blame the victim" message that chastises Canadians for not saving enough for retirement, despite the global financial downturn (Baldwin 2009; The Federal Task Force on Financial Literacy 2010) is ageist to the core because the message 130 years later is older people are not incapacitated, they are simply incompetent. The stereotype of incompetence is buttressed by the image depicted in the 2009 Federal Task Force on Financial Literacy (2010) that proclaimed, "many Canadians lack some or all of the skills, knowledge and confidence necessary to be financially literate" (2010, p. 4). This stereotype of the incompetent, illiterate older worker will lead potential retirees to the financial industry for paid advice. Financial institutions spend more than US$1.5 billion a year marketing their services to consumers in the United States and in Canada, through shared media. That amounts to nearly US$40 million a week of advertising dollars trying to catch the consumer's attention. Under these circumstances, confusion about retirement planning would not be uncommon, nor would the potential for "bad" retirement plans be small in a self-serving industry.

In the second instance, the form of private pension coverage began to shift from defined-benefit to defined-contribution plans, a transition witnessed in a number of countries (Baldwin and FitzGerald 2010). Although membership in defined-benefit plans still accounts for a majority of plan members, the trend to a declining portion of members in a defined-benefit plan is occurring in both the public and private sectors (Baldwin 2009). With the prevailing situation in Canada having begun in 2008, the financial circumstances of current and future retirees will become riskier, depending on the type of plan and investment as with defined-contribution plans (Beck 1992; Giddens 1990). It could be added that with the increase in defined-contribution plans, the responsibility for investment is now passed on to the individual and to his or her financial analyst or banker, if they can afford such a planner. There also is a serious conflict of interest in using financial planners because the financial industry may not be ready to provide the kind of information that people need, as opposed to the kind that sells products at high prices (Ambachtsheer 2008). With the universal pension long gone and the increasing attempt to replace social insurance with individualized market solutions, it seems that Canada is on the way back to earlier times, only with different negative images of older adults.

Third, as we have noted here, the situation is now changing in favour of later retirement. The changes in retirement policy per se have been few. For example, all provincial governments have recently eliminated contractual mandatory retirement at age 65 with the exception of federal employees, but even here the government plans to prohibit this practice. Pension plan changes are more in evidence. For example, recent changes have been made that support part-time work for some workers, since they are allowed to continue to accrue pension benefits if they work later; the age for contributing to RRSPs was raised to age 71 from age 69, and a tax credit was made available to encourage paid work for low-income earners aged 65 and older by reducing the disincentives to paid employment found in the Old Age Security/ Guaranteed Income Supplement programs.

In keeping with the plan to increase the age of retirement, several modifications to the Canada Pension Plan Act, recommended by the federal, provincial, and territorial governments, became law and will apply to those retiring in 2012. Those between the ages of 60 and 65 who plan to apply for a C/QPP pension early will have their benefits reduced. Workers age 65 and older can voluntarily elect to continue C/QPP participation, in which case regular employer contributions will also be required. In other words, early retirement will be penalized and later retirement rewarded. Most recently, an increase in the age of eligibility for the Old Age Security Pension to age 67 was announced by the Prime Minister while visiting another country, an announcement that surprised most Canadians since there had been no discussions of the matter. This back-door method to increase the age of retirement through pension policy has been suggested many times in the past to reduce the cost of pensions but has never really been successful (Brown 1995; Hicks 2003; Lam et al. 1997). The justifications are many: (a) gains in life expectancy requires resetting the retirement age (Denton and Spencer 1996, 2010; Hering and Klassen 2010); (b) a reduction in morbidity and mortality rules out a sickly labour force (Fries 1989); (c) retirement goes on too long (Hicks 2003); (d) generational equity demands later retirement, and (e) a later retirement age would be much less expensive (Hering and Klassen 2010). France, the United States, Germany, the United Kingdom, and Australia have already passed legislation that will, over time, raise the age of retirement. But the Canadian public has not had this discussion (Myles 2006) even though pension policy will shortly begin to encourage a later age of retirement. In Canada this is an example of social policy change by stealth where older people have not been consulted but rather have been pawns in the transformation of employment relations, supported by government interested in cost savings (McDonald and Donahue 2011).

Conclusions

The brief history of retirement has been provided to illustrate how retirement has been systematically ageist since it inception. Retirement has been used as a management tool by business and industry to manage the labour force of older adults over time, depending on whether it is financially fortuitous to business to remove or attach workers to their employment. Shifting ideologies about older adults that extend from negative to positive ageism have been used by business, government, the public and the media to support whatever justification is required at the time of retirement, with little thought about the harm perpetrated on older adults. The justifications span a wide array of images of older people from being dependent, vulnerable and frail through to job snatchers and greedy geezers that society has subscribed to en masse, including older adults themselves.

The argument that systematic ageism exemplified in retirement is a legitimate process as the result of the 'natural' functioning of competitive labour markets – deregulated "free markets" – is the greatest ageist idea of all. Free market economics is not self-regulating and is subject to government intervention as seen recently with the economic meltdown of 2008. Retirement is not a natural outcome of the functioning of the economy but is a handy social construction developed by business and industry. Neither is economics totally "scientific", no matter how strongly embraced by economists. The 'scientific' idea was only introduced in the last 40 years and it is easy to find many economic models that fail miserably at prediction and explanation (Cassidy 2009). Lastly, utopian economics and its offshoots are most decidedly not value free. The fuel behind free markets is human selfishness according to Adam Smith himself, wherein economic order will emerge as the unintended consequence of the actions of many people, each seeking their own self- interest (Cassidy 2009). The intervention of government in the economy found in any number of current political, economic and social media discourses is seen as simply unwise, if not morally wrong – a political philosophy laden with values (Cassidy 2009; The Economist 2010). An unstated value in the free market today is that older people are not esteemed or respected and as a result can be maneuvered at will to serve the forces of the market through retirement and pension polices.

How the next evolution of retirement will unfold is open to debate. What is significant is that older workers could possibly have more control in defining their own conditions in the economy simply on the basis of their large numbers, the fact that they will have to work to older ages and they no longer face mandatory retirement.

References

Abbott, M. G., Beach, C. M., Broadway, R. W., & MacKinnon, J. G. (Eds.). (2009). *Retirement policy issues in Canada*. Montreal: McGill-Queen's University Press.

Achenbaum, W. A. (1978). *Old age in a new land*. Baltimore: John Hopkins University Press.

Ambachtsheer, K. (2008). *The Canada Supplementary Pension Plan (CSPP): Towards an adequate, affordable pension for all Canadians*. Toronto: C. D. Howe Institute.

Atchley, R. C. (1976). *The sociology or retirement*. New York: Wiley.

Atchley, R. C. (1980). *"What happened to retirement planning in the 1970s?" Aging and retirement*. Beverly Hills: Sage.

Atchley, R. C. (1982). Retirement as a social institution. *Annual Review of Sociology, 8*, 263–287.

Atchley, R. C. (1993). Critical perspectives on retirement. In T. R. Cole, W. A. Achenbaum, P. L. Jakobi, & R. Kastenbaum (Eds.), *Voices and visions of aging: Toward a critical gerontology* (pp. 3–19). New York: Springer.

Baldwin, B. (2009). *Research study on the Canadian retirement income system*. Toronto: Ministry of Finance, Government of Ontario.

Baldwin, B., & FitzGerald, B. (2010). *Seeking certainty in uncertain times: A review of recent government-sponsored studies on the regulation of Canadian pension plans*. Toronto: C. D. Howe Institute.

Battle, K. (1997). Pension reform in Canada. *Canadian Journal on Aging, 16*(2), 519–552.

Beck, U. (1992). *Risk society: Towards a new modernity*. London: Sage.

Beland, D., & Myles, J. (2005). Stasis amidst change: Canadian pension reform in an age of retrenchment. In G. Bonoli & T. Shinkawa (Eds.), *Ageing and pension reform around the world* (pp. 252–272). Cheltenham: Edward Elgar Publishing.

Beland, D., & Myles, J. (2008). *Policy change in the Canadian welfare state: Comparing the Canada Pension Plan and Unemployment Insurance*. Hamilton: SEDAP.

Binstock, R. H. (1983). The Donald P. Kent memorial lecture. The aged as scapegoat. *Gerontologist, 23*(2), 136–143.

Binstock, R. H. (1994). Changing criteria in old-age programs: The introduction of economic status and need for services. *Gerontologist, 34*(6), 726–730 [Review].

Binstock, R. H. (2010). From compassionate ageism to intergenerational conflict? *Gerontologist, 50*(5), 574–585. doi:10.1093/geront/gnq056.

Brown, R. L. (1995). Achieving stability and equality and paygo financing. *Policy Options, 16*(6), 17–21.

Bryden, K. (1974). *Old age pensions and policy-making in Canada*. Montreal: McGill-Queen's University Press.

Butricia, A., Iams, H. M., Smith, K. E., & Toder, E. J. (2009). The disappearing defined benefit pension and its potential impact on the retirement incomes of boomers. *Social Security Bulletin, 69*(3), 1–27.

Bytheway, B. (1995). *Ageism*. Buckingham: Open University Press.

Canadian Institute of Actuaries. (1995). *Troubled tomorrows: The report of the Canadian Institute of Actuaries' task force on retirement savings*. Ottawa: Canadian Institute of Actuaries.

Cassidy, J. (2009). *How markets fail: The economics of rational irrationality*. New York: Farrar Straus Giroux.

Chappell, N. L., McDonald, L., & Stones, M. (2008). *Aging in contemporary Canada* (2nd ed.). Toronto: Pearson.
Commission, E. (2007). *Employment in Europe 2007*. Brussels: Employment Analysis Unit, European Commission.
Cumming, E. (1963). Further thoughts on the theory of disengagement. *International Social Science Bulletin, 15*, 377–393.
Denton, F. T., & Spencer, B. G. (1996). *Population aging and the maintenance of social support systems*. Hamilton: Program for Research on the Independence and Economic Security of the Older Population, McMaster University.
Denton, F. T., & Spencer, B. G. (2010). Chronic health conditions: Changing prevalence in an aging population and some implications for the delivery of health care services (Research Support, Non-U.S. Gov't). *Canadian Journal on Aging 29*(1):11–21. doi: 10.1017/S0714980809990390
Desmette, D., & Gaillard, M. (2008). When a "worker" becomes an "older worker": The effects of age-related social identity on attitudes towards retirement and work. *Career Development International, 13*, 168–185.
Donahue, W. T., & Tibbitts, C. (1957). *The new frontiers of aging*. Ann Arbor: University of Michigan Press.
Donahue, W., Orbach, H., & Pollak, O. (1960). Retirement: The emerging social pattern. In C. Tibbitts (Ed.), *Handbook of social gerontology* (pp. 330–406). Chicago: University of Chicago Press.
Duncan, C. (2001). Ageism, early exit and the rationality of age-based discrimination. In *Ageism in work and employment* (pp. 25–46). Aldershot: Ashgate.
Ekos Research Associates. (1995). *Rethinking government*. Ottawa: Ekos Research Associates Inc.
Estes, C., Swan, J. H., & Gerard, L. E. (1984). Dominant and competing paradigms in gerontology: Towards a political economy of ageing. *Ageing and Society, 2*, 151–164.
Financial Post. (1957, July 6). Retirement can kill you if leisure comes too fast.
Finnie, R., & Gray, D. (2011). *Labour-force participation of older displaced workers in Canada: Should I stay or should I go?* Montreal: Institute for Research on Public Policy (IRPP).
Fischer, D. H. (1978). *Growing old in America*. New York: Oxford University Press.
Foot, D. K., & Gibson, K. J. (1993). Population aging in the Canadian labour force: Changes and challenges. *Journal of Canadian Studies, 28*(1), 59–74.
Friedman, L. M. (2003). Age discrimination law: Some remarks on the American experience. In S. Fredman & S. Spencer (Eds.), *Age as an equality issue: Legal and policy perspectives* (pp. 175–194). Oxford: Hart Publishing.
Fries, J. F. (1989). *The compression of morbidity: Near or far?* [Clinical trial randomized controlled trial research support, U.S. Gov't, P.H.S. Review]. *Milbank Quarterly, 67*(2), 208–232.
Giddens, A. (1990). *The consequences of modernity*. Oxford: Polity Press.
Gillin, C. T., & Klassen, T. R. (1995). Age discrimination and early retirement policies: a comparison of labor market regulation in Canada and the United States [Comparative study review]. *Journal of Aging & Social Policy, 7*(1), 85–102. doi:10.1300/J031v07n01_06.
Gillin, C. T., & Klassen, T. R. (2005). The shifting judicial foundation of legalized age discrimination. In C. T. Gillin, D. MacGregor, & T. R. Klassen (Eds.), *Time's up: Mandatory retirement in Canada* (pp. 45–73). Toronto: James Lorimer & Company Ltd.
Globe and Mail. (1995, February 25). *Canadians want strong active government*, p. A6.
Globe and Mail. (2009). *Retirement dreams under siege*. Retrieved from http://www.theglobeandmail.com/report-on-business/retirement/
Gougeon, P. (2009). *Shifting pensions*. Ottawa: Statistics Canada.
Graebner, W. (1980). *A history of retirement*. New Haven: Yale University Press.
Grattan, P. (2002). Age discrimination in employment. In Help the Aged (Ed.), *Age discrimination in public policy: A review of evidence* (pp. 33–35). London: Help the Aged.
Guillemard, A. M. (Ed.). (1983). *Old age and the welfare state*. London: Sage.
Haber, C. (1978). Mandatory retirement in nineteenth-century America: The conceptual basis for a new work cycle. *Journal of Social History, 12*(1), 77–96.
Haber, C., & Gratton, B. (1994). *Old age and the search for security: An American social history*. Bloomington: Indiana University Press.

Halliwell, C. (2009). Retirement in Canada: Some trends...Some issues. In M. G. Abbott, C. M. Beach, R. W. Broadway, & J. G. MacKinnon (Eds.), *Retirement policy issues in Canada* (pp. 49–64). Montreal: McGill-Queen's University Press.

Hardy, M. A., Hazelrigg, L. E., & Quadagno, J. S. (1996). *Ending a career in the auto industry: 30 and out*. New York: Plenum Press.

Hering, M., & Klassen, T. R. (2010). *Why we need to raise the retirement age*. Toronto: The Mowat Centre for Policy Innovation.

Hicks, P. (2003). New policy research on population aging and life-course flexibility. *Horizons, 6*, 3–6.

Hicks, P., Halliwell, C., Hebert, B. P., & Lenjosek, G. (2005). Some perspectives on changing the pension system. *Canadian Public Policy, 34*(1), 15–20.

House of Commons Debates. (1921). *House of commons debates of the dominion of Canada fifth session: Thirteenth parliament*. Ottawa: F. A. Ackland King's Printer.

Ibbott, P., Kerr, D., & Beaujot, R. (2006). Probing the future of mandatory retirement in Canada [Review]. *Canadian Journal on Aging, 25*(2), 161–178.

International Longevity Center. (2006). *Ageism in America*. New York: The Anti-Ageism Taskforce, International Longevity Center.

Klassen, T. R., & Gillin, C. T. (1999). The heavy hand of the law: The Canadian Supreme Court and mandatory retirement. *Canadian Journal on Aging, 18*(2), 259–276.

Kohli, M., & Rein, M. (1991). The changing balance of work and retirement. In M. Kohli, M. Rein, A. M. Guillemard, & H. van Gunsteren (Eds.), *Time for retirement: Comparative studies of early exit from the labour force* (pp. 1–35). Cambridge: Cambridge University Press.

Labour Gazette. (1957). *Compulsory retirement said cruel, wasteful*, pp. 57, 933.

Laczko, F., Dale, A., Arber, S., & Gilbert, G. N. (1988). Early retirement in a period of high unemployment. *Journal of Social Policy, 17*(3), 313–333.

Lam, N., Cutt, J., & Prince, M. J. (1997). The Canada pension plan: Retrospect and prospect. In K. Banting & R. Boadway (Eds.), *Reform of the retirement income system: International and Canadian perspectives* (pp. 105–134). Kingston: Queen's University, School of Policy Studies.

Levy, B. R. (2001). Eradication of ageism requires addressing the enemy within [Comment]. *Gerontologist, 41*(5), 578–579; discussion 580.

Little, B. (2008). *Fixing the future: How Canada's usually fractious governments worked together to rescue the Canada Pension Plan*. Toronto: University of Toronto Press.

Longino, C. F., Jr. (2005). The future of ageism: Baby boomers at the doorstep. *Generations, 29*(3), 79–83.

Lowe, G. S. (1991). Retirement attitudes, plans and behaviour. *Perspectives, 3*(3), 8–17.

Macleans. (1961). *A case for retiring but at 50, a case: For retiring not at 50, a case: For retiring (not quitting) at 50*. November 18, 71.

Macnicol, J. (1998). *The politics of retirement in Britain, 1878–1948*. Cambridge: Cambridge University Press.

Macnicol, J. (2006). *Age discrimination: An historical and contemporary analysis*. Cambridge: Cambridge University.

Macnicol, J. (2008). Older men and work in the twenty-first century: What can the history of retirement tell us? *Journal of Social Policy, 37*(4), 579–595.

Macnicol, J. (2010). *Ageism and age discrimination: Some analytical issues*. London: International Longevity Centre.

Marshall, V. W., & Marshall, J. G. (2003). Ageing and work in Canada: Firm policies. *The Geneva Papers on Risk and Insurance, 28*(4), 625–639.

McCann, R. M., & Giles, H. (2002). Ageism and the workplace: A communication perspective. In T. D. Nelson (Ed.), *Ageism* (pp. 163–199). Cambridge: MIT Press.

McDonald, L. (1995). Retirement for the rich and retirement for the poor: From social security to social welfare. Editorial. *Canadian Journal on Aging, 14*(3), 447.

McDonald, L. (1997). The link between social research and social policy options: Reverse retirement as a case in point. *Canadian Journal on Aging/Public Policy, 23*(S1), 90–113.

McDonald, L. (2002). *The invisible retirement of women*. Hamilton: SEDAP Research Program.
McDonald, L. (2006). Gendered retirement: The welfare of women and the "new" retirement. In L. Stone (Ed.), *New frontiers of research on retirement* (pp. 137–164). Ottawa: Statistics Canada.
McDonald, L. (2011). Elder abuse and neglect in Canada: The glass is still half full. *Canadian Journal on Aging, 30*(3), 1–30.
McDonald, L., & Donahue, P. (2011). Retirement lost? *Canadian Journal on Aging, 30*(3), 401–422. doi:10.1017/S0714980811000298.
McDonald, L., & Wanner, R. A. (1990). *Retirement in Canada*. Toronto: Butterworths.
McDonald, L., Donahue, P., & Marshall, V. (2000). The economic consequences of unexpected early retirement. In F. T. Denton, D. Fretz, & B. G. Spencer (Eds.), *Independence and economic security in old age* (pp. 267–292). Vancouver: University of British Columbia Press.
McVittie, C., McKinlay, A., & Widdicombe, S. (2003). Committed to (un)equal opportunities: 'New ageism' and the older worker [Research support, non-U.S. Gov't]. *British Journal of Social Psychology, 42*(Pt 4), 595–612. doi:10.1348/014466603322595293.
Meier, E. L. (1986). *Employment experience and income of older women*. Washington, DC: American Association of Retired Persons.
Midgley, J. (1997). *Social welfare in global context*. Thousand Oaks: Sage.
Mintz, J. M. (2009). *Summary report on retirement income adequacy research*. Ottawa: Department of Finance Canada.
Morton, D., & McCallum, M. E. (1988). *Superannuation to indexation: Employment pension plans*. Toronto: Queen's Printer for Ontario.
Moussaly, K. (2010). *Participation in private retirement savings plans, 1997 to 2008*. Ottawa: Authority of the Minister, Statistics Canada.
Myles, J. (2006). From pension policy to retirement policy: Towards a new social agenda? In L. Stone (Ed.), *New Frontiers of research on retirement* (pp. 65–82). Ottawa: Statistics Canada.
National Council of Welfare. (1984). *Sixty five and older*. Ottawa: Minister of Supply and Services Canada.
OECD. (1988). *Aging populations: The social policy implications*. Paris: Organisation for Economic Co-operation and Development (OECD).
OECD. (1995). *The transition from work to retirement*. Paris: Organisation for Economic Co-operation and Development (OECD).
OECD. (2009). *OECD private pensions outlook 2008*. Paris: Organisation for Economic Co-Operation and Development (OECD).
OECD. (2010). *Pensions in France and abroad: Seven key indicators*. Paris: Organisation for Economic Co-Operation and Development (OECD).
Pakans, A. (1989). Stepping down in former times: A comparative assessment of retirement in traditional Europe. In D. I. Kertzer & K. W. Schaie (Eds.), *Age structuring in comparative perspective* (pp. 175–195). Hillsdale: Lawrence Erlbaum Associates.
Palmore, E. (1990). *Ageism: Negative and positive*. New York: Springer.
Palmore, E. (2005). Three decades of research on ageism. *Generations, 29*(3), 87–90.
Parliament of Canada. (1926–27). *Debates of the Senate of Canada*. Ottawa: Parliament of Canada.
Parsons, T. (1942). Age and sex in the social structure of the United States. *American Sociological Review, 7*, 604–616.
Peel, N. M., McClure, R. J., & Bartlett, H. P. (2005). Behavioral determinants of healthy aging [Research Support, Non-U.S. Gov't Review]. *American Journal of Preventive Medicine, 28*(3), 298–304. doi:10.1016/j.amepre.2004.12.002.
Pentland, H. C. (1981). *Labour and capital in Canada, 1650–1860*. Toronto: James Loremont.
Phillipson, C. (1982). *Capitalism and the construction of old age*. London: Macmillan.
Phillipson, C. (1999). The social construction of retirement: Perspectives from critical theory and political economy. In M. Minkler & C. L. Estes (Eds.), *Critical gerontology: Perspectives from political and moral economy* (pp. 315–318). New York: Baywood.
Phillipson, C. (2004). Work and retirement transitions: Changing sociological and social policy contexts. *Social Policy & Society, 3*(2), 155–162.

Phillipson, C., & Smith, A. (2005). *Extending working life: A review of the research literature*. Norwich: Department for Work and Pensions.
Pignal, J., Arrowsmith, S., & Ness, A. (2010). *First results from the survey of older workers, 2008*. Ottawa: Special Surveys Division, Statistics Canada.
Platman, K., & Taylor, P. (2004). Age, employment and policy. *Social Policy & Society, 3*(2), 143–144.
Quinn, J. F., Burkhauser, R. V., & Myers, D. L. (1990). *Passing the torch: The influence of economic incentives on work and retirement*. Kalamazoo: W. E. Upjohn Institute for Employment Research.
Rix, S. E. (2011). *Recovering from the Great Depression: Long struggle ahead for older Americans*. Washington, DC: AARP Public Policy Institute.
Robertson, A. (1991). The politics of Alzheimer's disease: A case study in apocalyptic demogrpahy. In M. Minkler & C. Estes (Eds.), *Critical perspectives on aging: The political economy and moral economy of growing old* (pp. 135–150). Amityville: Baywood.
Rowe, G., & Nguyen, H. (2003). Older workers and the labour market. *Perspective on Labour and Income, 15*(1), 55–58.
Rozanova, J., Northcott, H., & McDaniel, S. (2006). Seniors and portrayals of intra-generational and inter-generational inequality in The Globe and Mail. *Canadian Journal on Aging, 25*(4), 373–386.
Sass, S. A., Monk, C., & Haverstick, K. (2010). *Workers' response to the market crash: Save more, work more?* Chestnut Hill: Center for Retirement Research at Boston College.
Schellenberg, G. (1994). *The road to retirement: Demographic and economic changes in the 90's*. Ottawa: Canadian Council on Social Development.
Schellenberg, G. (2004). *2003 general social survey on social engagement, cycle 17: An overview of findings*. Ottawa: Statistics Canada.
Schellenberg, G., Turcotte, M., & Ram, B. (2005). Preparing for retirement. *Canadian Social Trends, 78*, 8–11.
Simpson, I. H. (1975). Review symposium on work in America. *Sociology of Work and Occupations, 2*, 182–187.
Siroonian, J. (1993). *Work arrangements*. Ottawa: Statistics Canada.
Snell, J. G. (1996). *The citizen's wage: The state and the elderly in Canada, 1900–1951*. Toronto: University of Toronto Press.
Taylor, F. (1947). *Scientific management*. New York: Harper.
Teeple, G. (1995). *Globalization and the decline of social reform*. Toronto: Garamond Press.
The Daily. (2010). Pension plans in Canada. Retrieved from http://www.statcan.gc.ca/daily-quotidien/100525/dq100525c-eng.htm
The Economist. (2010). 49th parallels: Stephen Harper has imported American Republicanism just when Barack Obama has adopted some Canadian policies. Is there still a North American divide? Retrieved from http://www.economist.com/node/16377327
The Federal Task Force on Financial Literacy. (2010). *What we heard: A summary of public consultations* (M. o. Finance, Trans.). Ottawa: Minister of Finance.
The Globe. (1905, February 24). Man's end is at forty.
Tompa, E. (1999). *Transition to retirement: Determinants of age of social security take up*. Hamilton: SEDAP, McMaster University.
Townsend, P. (1981). The structured dependency of the elderly: Creation of a social policy in the 20th century. *Ageing and Society, 1*, 5–28.
Townsend, P. (2006). Policies for the aged in the 21st century: More "structured dependency" or the realisation of the human rights? *Ageing and Society, 26*(2), 161–179.
Union, N. (2007). *Pensions backgrounder #2: A brief history of pensions in Canada*. Nepean: National Union of Public and General Employees.
Walker, A. (1990). The benefits of old age? Age discrimination and social security. In E. McEwen (Ed.), *Age: The unrecognised discrimination*. London: Age Concern.
Walker, A. (2000). Public policy and the construction of old age in Europe. *Gerontologist, 40*(3), 304–308.

Walker, A. (2006). Active ageing in employment: Its meaning and potential. *Asia-Pacific Review, 13*(1), 78–93.
Wallace, E. (1950). The origin of the Social Welfare State in Canada, 1867–1900. *Canadian Journal of Economics and Political Science, 16*(3), 383–393.
Wang, M., & Shultz, K. S. (2010). Employee retirement: A review and recommendations for future investigation. *Journal of Management, 36*, 172–206.
Weaver, R. K. (2004). Pension reform in Canada: Lessons for the United States. *Ohio State Law Journal, 65*(1), 44–74.
Workman, T. (1996). *Banking on deception: The discourse of fiscal crisis*. Halifax: Fernwood Publishing.

Part III
Future Solutions

Chapter 6
Creating Supportive Workplace Environments for Older Workers

Elizabeth Ciampa and Roslyn Chernesky

The U.S. population is growing older. Due to lower birth and mortality rates, experts predict that persons over 65 years of age will increase in coming years, both in absolute numbers and as a percentage of the population (Moody and Sasser 2012; Marshall and Altpeter 2005). The number of those 65 and older is expected to double from 39.6 million in 2009 to 72.1 million in 2030 (Administration on Aging 2011; Sanders et al. 2008).

Retirement policies and labor force participation have changed in line with the aging of the U.S. population. Recent years have seen declining birth rates after the baby boom generation born from 1946 to 1964, increasing retirement ages, increasing financial incentives for delayed retirement, and abolishment of Social Security earnings tests for workers between full retirement age and 70 years old (Gustman and Steinmeier 2009; Klay and Steen 2008). As a result, the full-time labor force participation of those aged 65–67 years of age has increased by over 9% between 1992 and 2004.

As labor force participation by older workers increases, there will be fewer 16–24 year olds to replace the millions of workers reaching retirement age (Leonesio et al. 2012). Between 2008 and 2018 it is projected that the number of workers aged 55 and older will grow by nearly 43% (Bureau of Labor Statistics 2009). The rate of unemployment for older adults is at a record high of 7.3%, with 2.2 million of the 14.9 million unemployed individuals aged 55 years of age or older (Rich 2010). Coupled with the aging of the population, increased global competition, business deregulation, technological changes, and corporate mergers and acquisitions, changes in the composition of employment and the workforce are taking place (Sparks et al. 2001).

E. Ciampa, M.S. (✉) • R. Chernesky, D.S.W. (✉)
Department of Social Services, Fordham University, New York, NY, USA
e-mail: Ciampa.liz@gmail.com; Chernesky@fordham.edu

P. Brownell and J.J. Kelly (eds.), *Ageism and Mistreatment of Older Workers: Current Reality, Future Solutions*, DOI 10.1007/978-94-007-5521-5_6,

Influence of Demographic Shifts on the Workplace

These aging population trends will inevitably increase the importance of organizational workplace structures and polices on older worker employment (Gibson et al. 2010; Klay and Steen 2008; Neumark 2008). Lower labor force participation by younger workers necessitates that employers turn to older workers to ensure companies' economic and financial stability. Organizational structures and public policy must reflect this change. As older workers comprise an increasing share of the nation's labor force, there is a greater need to consider how older workers fare in their workplaces.

These demographic shifts in favor of older populations and workforces are also taking place globally. These shifts in the world's aging population challenge traditional retirement structures, employment and social policies, labor market trends, global workforces, organizational work environments, and social insurance programs (AARP 2007). Organizational practice strategies aimed at developing and maintaining age-friendly work environments are imperative. A recent survey that examined multi-stakeholder views on recent demographic changes to the workforce across different countries revealed vast challenges and opportunities for governments, individuals and employers. Many older workers want to, and must, remain productive members of the workforce. At the same time, employers will have to rely more heavily on older workers. With an aging workforce it will be necessary to develop intergenerational and cohesive work environments (AARP 2007).

A number of companies have recognized this imperative and have implemented flexible work policies, created attractive opportunities to promote and maintain employment, and redesigned workplaces to accommodate older workers. Many businesses have devised strategies for the retention of older workers. These strategies allow companies to adapt to changing labor market demands. In addition, they can increase U.S. economic output and tax revenues (Eyster et al. 2008). Although many company-sponsored retirement plans provide large worker incentives to retire, employers should nevertheless evaluate the personal circumstances of individual older employees. For example: Are these older workers willing to continue to work? What happens when they return to work? Do they find their jobs fulfilling? and, Will they be financially secure after retirement?

Impact of Ageist Beliefs in Society and the Workplace

Contemporary society is characterized by stereotypes that associate old age with physical and mental decline. Changes to organizational culture and climate that might help older workers are often limited due to a lingering negative perception of age. Common misconceptions of older workers, although inaccurate, pervade today's workplace. Such misconceptions include the beliefs that older workers: take more time off; have decreased performance; have more accidents; are less adaptable;

are merely waiting to retire; take jobs away from younger workers; and are incapable of learning new things (AARP 2007). Additional misconceptions are that older workers experience greater fatigue; are more resistant to change; unenthusiastic; less knowledgeable; less interested in receiving training; and less willing to gain new knowledge when compared to younger workers (Bulter 2010). These ageist stereotypes deny older workers opportunities and resources. Thus, policy analysts and employers must be aware of these stereotypes in order to understand the labor force participation of older workers.

These stereotypes are rooted in long-standing and pervasive public attitudes. According to Clark (1997), the term "old age" has held various meanings over time based on what he calls the *decline model*. The decline model views old age as a time of decay, decline and disengagement. Old age is viewed as a period of life in which older persons lose their health, power, authority, credibility, vital energy and morality. It is perceived as a period of physical and mental decline (Clark 1997). Due to a loss of intellect and understanding, individuals are assumed to no longer have interests in work, events, society or people. Others assume that such lack of vital energy and mental capacity dictates a withdrawal and separation from society in old age, regardless of health. Many workers believe they should retire before old age, and therefore many retire long before they become incapable of working. Schrank and Waring (1989) argued that Americans are ambivalent towards older workers: on one hand, they honor them for being productive into old age; on the other hand, they believe that older workers are disengaged, dependent, and unproductive. These beliefs are reflected in workplace policies and practices.

This chapter considers the impact of these stereotypes on older adults' workforce participation, productivity and satisfaction. It offers practice strategies for administrators and managers to support older workers and foster age-friendly workplace environments.

Age-related organizational policies are inconsistent and tend to revolve around the false notion that older workers cost more and are less productive than younger workers (Posthuma and Campion 2009). The realities of older workers, especially those from the baby boomer generation, dramatically contradict this misconception. According to Gibson et al. (2010), baby boomers are characteristically loyal and competitive workers and hold values influenced by the economic prosperity following World War II and the Vietnam War. These workers are likely to be in better health, better educated, trained in technologies, and able to compensate for the chronic illness and disabilities common among older workers in the past. Compared to younger workers, older workers are often sick less often, get hurt less often, have fewer absences for dependent-care obligations and fewer distractions that interfere with work-related commitments. Organizations often want to rid their companies of older workers in an effort to obtain "new blood" and therefore older workers are encouraged to leave for no legitimate reason. It is ageism, rather than labor cost and performance considerations, that is the reason corporations force out older workers. Despite a lack of evidence that younger workers are more intelligent, creative, dedicated or productive, they are often favored (Posthuma and Campion 2009).

Workplace Environments

Historically, policy makers have focused on opening doors so that older adults can enter into employment. Much of the literature has therefore focused on policies and practices that create barriers for older workers seeking jobs (Gibson et al. 2010; Posthuma and Campion 2009; Stark 2009). Today policy makers are more interested in what happens to older adults in the workforce. As evidenced in this book, ageism, abuse and harassment in the workplace are realities. There is little emphasis, however, on subtle and indirect ways older adults experience obstacles to full participation in their workplaces. These often result from unintentional practices that reflect ageist stereotyping. Some are by-products of the aging process itself.

A workplace environment reflects the beliefs, vision, and leadership style of top management; these determine workplace practices. Organizational leaders influence workers' commitment to their jobs and to their organizations, worker satisfaction, and productivity. In human resources departments, the attitudes of personnel to their work situation and to their coworkers also affect how staff relate to clients or customers and to their effectiveness in serving them (Garrow and Hasenfeld 2010).

The quality of the workplace environment is determined by the extent to which older workers are valued as contributors and are able to experience job satisfaction. The climate and culture of organizations can create a workplace environment that either is supportive and satisfying for older workers or is inhospitable and hostile for them (Chernesky 1998; Denison 1996). *Organizational culture* refers to the deep structure of organizations, rooted in the values, beliefs, and assumptions of organizational members. It is established through socialization to a variety of identity groups that converge in the workplace. Interaction reproduces a symbolic world that gives culture both stability and, at the same time, a precarious and fragile nature rooted in the dependence of the system on individual cognition and action. The culture of an organization is reflected in the pattern of beliefs, values, and expectations shared by organization members that establish norms and rules governing acceptable and unacceptable employee behaviors. The standards against which behavior is evaluated are that of the majority culture, and reflect the attitude and values of those who hold positions of authority and power. Persons who are perceived as different or outsiders may not measure up to the standards, and therefore may not be allowed to participate fully in the organization and reap its benefits (Chernesky 2005).

Some workplaces are hostile and offensive environments. In such workplaces, discrimination, harassment, slurs, jokes, and unintentional acts that reinforce biases, differences, and disparities among individuals are tolerated. These workplace characteristics pollute the work environment, disrupt productive work, and lead to burnout (Chernesky 1998). An organization's culture can be tolerant of behaviors that discriminate or harass, thus making the workplace inhospitable and offensive to workers who are not members of the majority group. Older and disabled workers may face hostile workplace environments due to discrimination based on age and disability. These forms of workplace discrimination are widespread (Rampell 2011a). The situation today is complicated by an increasing number of on-the-job discrimination complaints, fueled perhaps by employee anxiety and insecurity in the workplace due to the tight economy with its resulting real or potential layoffs.

Women and minority older workers face additional challenges in the workplace. They may be the targets of *micro inequities*. Micro inequities or micro aggressions are unjust actions taken against individuals based on stereotypes related to race or gender. They are often unintentional acts that reinforce biases, differences, and disparities among individuals in organizations. They marginalize targeted workers who are then not accepted and acknowledged by their co-workers, supervisors and employers (Laporte 1991). They are a form of subtle discrimination, either conscious or unconscious, and are common in the workplace. They are usually so subtle that it is difficult to know how to counter them. Examples include: a woman who speaks up at a meeting and later finds that her remarks were attributed to the man seated next to her; and the African-American professional who is mistaken for a support person instead of a higher-level employee.

Many situations exist where slurs, jokes and epithets, although not threatening, pollute the work environment. Humor and joking can relieve tension or frustration, reduce boredom, reinforce status levels, establish roles, and provide a means to express feelings where there are no other socially acceptable outlets. However, if these insults are frequent, or implicitly or explicitly condoned by management, they are likely to violate Title VII or other fair employment laws (Chernesky 2005; Laporte 1991). Organizations with a high frequency of such insults may be characterized as *toxic organizations*.

Organizational toxicity is defined as "the outcome of emotionally insensitive attitudes and actions of managers and the practices of their companies… [that act] as a noxious substance, draining vitality from individuals and [the]…..entire organization" (Frost 2003, p. 13). Toxicity is present to some extent in all organizations. If ignored, it can significantly disrupt innovative and productive work. Whether introduced into an organization from outside or within, if it is not handled well, managers and workers can "lose heart", stop thinking creatively, and become anxious and fearful (Bailey et al. 2008). Toxic workplaces tolerate micro inequities and micro-aggressions (Chernesky 2005) that allow tensions and conflicts among workers to dominate, and make it possible for toxic elements to fester (Frost 2003). In such poorly-managed workplace environments, workers may become detached and experience burnout (Maslach 1998). Unhappy and stressed workers are also at higher risk of health problems (Sutton 2010). Such toxic environments are not only damaging to employees but also increase costs and undermine company performance (Sutton 2010).

Organizational climate, in contrast to organizational culture, "is widely defined in terms of employees' perceptions of their work environment" (Glisson 2000, p. 196). The climate of any given organization is an aggregation, a synergistic combination, of individual psychological climates. Organizational climate is important because of its impact on workers' behavior and attitudes. It also can influence organizational effectiveness and efficiency. How supportive an environment is perceived to be by workers will determine their productivity and the quality of their performance. Managers in supportive workplace environments recognize that the conditions of work have a profound impact on workers' capacity to fulfill their own needs (including basic security, relational, self-fulfillment and actualization); on their attitudes about work and co-workers; and ultimately, on how they perform their jobs (Garrow and Hasenfeld 2010).

Culture and climate are powerful because they affect relationships among workers, and interactions between workers and to whom they provide services, clients or customers (Hemmelgarn et al. 2010; Glisson and Hemmelgarn 1998). Rampell (2011b) found that older workers think of younger workers as lazy slackers. Older workers, unlike younger ones, have a strong sense of separate spheres for work and play, and preference for working independently rather than turning to colleagues for help. Thus, a cultural clash between younger and older workers is not unusual (Rampell 2011b). Older workers with younger co-workers, supervisors, or bosses may perceive them as "kids" who feel "entitled" and who are not serious about their jobs. Unless diffused, this causes intergenerational conflict and tension (Daily 2011). Even perceptions of courtesy, civility and rude behavior are likely to have a generational dimension that can affect worker relationships (Tugend 2010).

Practice Strategies for Age-Friendly and Abuse-Free Workplace Environments

Supportive workplaces promote worker motivation and satisfaction. Managers should ensure that the organizational climate and culture support older workers and enable their companies to reap the benefits of employing older workers. Through the use of proper strategies, managers can reduce or prevent toxic work environments. They can augment worker strengths, protect workers, and amplify worker resilience to adverse organizational conditions (Zunz and Chernesky 2000). In the following discussion, the authors describe a number of such strategies.

AARP Initiatives

As discussed earlier, demographic shifts are producing profound changes to workplace environments. As the proportion of younger workers declines, employers increasingly want to attract and retain mature employees. AARP has helped in this regard by advocating for older workers to remain productive and powerful members of the workforce. In turn, older worker labor force participation has yielded increased economic growth and social progress (AARP 2007).

AARP initiatives have been designed to assist employers in evaluating their workforce needs and developing effective aging workforce strategies. These strategies highlight the importance of developing age-friendly workplace environments. Through the International Innovative Employer Award, AARP has recognized 15 non-U.S.-based employers for implementing best practices to attract and retain older workers (AARP 2011a). Award recipients are acknowledged for enhancing the retention of older workers through policies such as: flexible work arrangements; lifelong learning, training, and recruitment; and the promotion of health and diversity (AARP 2011a).

In 2011, the title of this award program was changed to the AARP Best Employers for Workers over 50 Award. The award recognized 15 non-U.S.-based employers whose workforce and/or human resource practices were successful in addressing relevant issues pertinent to workers aged 50 and older, and created roadmaps for attracting and retaining tomorrow's aging workforce (AARP 2011b). The award has both U.S. and international components and is given on a bi-annual basis in an effort to allow employees time to implement and test innovative strategies to meet the needs of the aging workforce.

Other innovative company practices include: pre-retirement seminars; time off for family care giving; employer-provided elder care services; corporate retirement housing options; and self-confidence and employability training. In an effort to retain older workers, some employers have worked to identify factors that lead to workers' decisions to retire. They have identified skills transferable across job categories; targeted the health needs of mature workers; provided on-site technical assistance; instituted workforce literacy training; and provided flexible work and retirement options (Foggs 2001; NCOA 1991). Some employers have also designed programs to teach employees about generational differences in real-world situations to create respect in the workplace so that people from all generations can work together efficiently. Employers increasingly recognize the importance of a hospitable workplace climate for their older workers and ensure that all practices are age-neutral (Fogg 2001; NCOA 1991). A handful of organizations have begun to ignore age in identifying new work opportunities and measuring performance; instead, they have focused on job competence and performance (Standing and Baume 2001).

AARP (2011a, b) selected award winners who have successfully implemented such strategies through the 2010 AARP International Innovative Employer Award and the 2011 Best Employers for Workers Over 50. Awards winners included:

Adecco Employment Services Limited: Toronto, Canada
Winning Years: 2009, 2010
Adecco implemented a national hiring strategy that actively targets job seekers aged 50 and older through collaboration with its "Established Worker Career-Connections Program" and the Canadian Association for Retired Persons (CARP). This strategy ensures the hiring process is free from age discrimination, even providing paper applications to those not comfortable using computers. Those candidates re-entering the workforce are offered training on interviewing practices and techniques to market their skills.

BMW Group: Munich, Germany
Winning Year: 2011
BMW Group created a program including measures addressing demographic changes in the workplace to ensure a well-balanced age structure workforce. This program includes initiatives related to health management, training and knowledge management, work environment, and personalized retirement models.

BT Telecommunications: London, United Kingdom
Winning Years: 2008, 2009, 2010
Through its "Achieving the Balance" program, BT helps older employees transition from full-time employment to retirement. Through its "Career-Life Planning Tool", BT helps employees in developing their careers at every stage of employment.

Bundesagentur für Arbeit: Nürnberg, Germany
Winning Years: 2010, 2011
Through its flexible work schedules, Bundesagentur für Arbeit (BA) offers over 500 work schedules to accommodate the specific needs of their employees. Long-term working time accounts are also offered, allowing employees to extend their leaves of absence for issues such as caring for family members and transitioning into retirement. BA also employs a comprehensive health promotion strategy including re-integration programs, which allow employees to return to the workplace following a long-term sick leave.

Centrica plc: Windsor, United Kingdom
Winning Years: 2008, 2009, 2010, 2011
Through its "Age Action Group", Centrica aims to create an age-diverse workforce through having managers from each of the different business units create a plan addressing the company's aging workforce. Additionally, Centrica created an "Age Awareness" e-learning package that raises awareness about age-diversity for managers and employees.

Deutsche Lufthansa AG: Hesse, Germany
Winning Year: 2010
A variety of flexible work arrangements and continuous learning and training are offered to employees, allowing for part-time schedules, job sharing, telecommuting and e-learning. Individualized elder care solutions are offered to employees with caregiving responsibilities as well as counseling services for those facing personal challenges and crises. Partial retirement and intermediate pensions are provided, which cover the years between stopping flying and retirement.

Daikin Industries: Osaka, Japan
Winning Year: 2011
Opportunities for individual capacity building, personal development, and flexible work schedules are offered to employees. Extended family caregiving leave options and rehiring of those wishing to work past the statutory retirement age are offered to all employees.

DB Services: Berlin, Germany
Winning Year: 2011
Older workers are promoted through a training camp, the DB Services Academy, which helps older job applicants and the long-term unemployed re-enter the workforce. Those employees who are unable to continue to work in their physically demanding positions are referred to their administrative business unit, designed to employ these workers.

DSW 21: Dortmund, Germany

Winning Year: 2011

Retraining and alternate position options are offered to employees who no longer have the capacity to perform their original jobs.

Elkerliek Hospital: Helmond, Netherlands

Winning Year: 2011

Workforce management is based on the Work Ability Index (WAI) analysis, which focuses on creating new positions that fit the needs of an employee with a disability. Risk profiles are also created for each employee in order to manage physical and psychological stress.

Galeria Kaufhof GmbH: Cologne, Germany

Winning Year: 2010

A "Fit for the Future" program focuses on individual training for older employees, which works to enhance qualifications, employability and confidence. "Health days" are offered to workers over age 55 to promote the company's holistic health program. Additionally, a Demography Officer works to ensure all employment policies are congruent with the latest demographic analysis.

Guernsey Post: Guernsey, United Kingdom

Winning Year: 2010

The changing needs of older workers are accommodated in reploying staff to positions that are less physically demanding. Pre-retirement training, occupational health promotion, and in-house physiotherapy are offered to mature workers, although Guernsey Post has no compulsory retirement age.

Hansgrohe AG: Schiltach, Germany

Winning Year: 2010

Diversity in work environments is promoted through Project MUMM, which was designed to increase the motivation and participation of mature workers through computer training, flexible work policies, health promotion, and alternative working hours.

Henkel AG & Co: Düsseldorf, Germany

Winning Year: 2010

Flexible work arrangements are offered and those approaching retirement are given a 5-year, part-time schedule, which cuts their hours up to 50% but allows their social contributions to be paid on a full-time basis. Health promotion events are held regularly and generational knowledge transfers are supported through multigenerational roundtable discussions.

Jena-Optronik, Gmbh: Jena, Germany

Winning Year: 2011

A "55plus" comprehensive framework promotes a positive internal working atmosphere and the employability of older workers and includes features such as flexible accounts, training modules, and active promotion of mixed-age work teams.

Lam Soon Edible Oils: Shah Alam, Malaysia
Winning Year: 2011
Older workers are designated as subject matter expects and serve as facilitators and trainers. Flexible work arrangements and the ability to remain employed beyond statutory retirement age are offered to employees.

Marks and Spencer plc: London, England
Winning Year: 2011
Partial retirement schemes, health promotion initiatives and discussion boards contribute to the retention of older workers.

National Australia Bank Limited: Melbourne, Australia
Wining Year: 2011
The MyFuture initiative offers education on flexibility, work-life balance and managing personal change to enable mature workers to extend their careers. Health promotion initiatives such as check-ups, interactive health portals, and confidential support services help retain older workers.

National Environment Agency: Singapore
Winning Year: 2011
Older workers serve as mentors to younger staff and instructors for internal training programs. Experienced older workers are given the option to deploy to temporary overseas projects as consultants.

National University Hospital: Singapore
Winning Year: 2010
The "Active Edge" program aims to increase the mature workforce. Pre-retirement planning and counseling are offered to employees at least 3 years prior to age 62. A variety of re-employment opportunities and alternative work arrangements and benefits are then offered at age 62.

PUB: Singapore
Winning Year: 2010
Flexible work options, paid caregiver days, pre-retirement consulting, and health promotion workshops aim to maintain a mature workforce. Re-employment of workers passed the statutory retirement age is stressed to allow workers to continue their same job or be hired for a new one.

Raffles Institution: Singapore
Winning Year: 2011
A reimbursement of up to $1,500 is provided for spending on anything that promotes health or work-life balance. Alternative work options are provided to help employees' manage work and family commitments.

Salzburg AG fur Engergie, Verkehr und Telekommunikation: Salzburg, Austria
Winning Years: 2010, 2011
Life-long learning and training and health promotion are offered to employees via the "Genera Generation Management" and "Giving Energy-Living Energy" programs. Needs and concerns are assessed through employee surveys and a number of courses are offered to address these needs.

SICK AG, Industrial Equipment/Commercial Machinery: Waldkirch, Germany
Winning Year: 2010
Mixed-age training groups and project teams have older workers serve as mentors for younger employees. Healthy aging in the workplace is stressed through an emphasis on flexible work options.

Sozial-Holding der Stadt Mönchengladbach GmbH: Mönchengladbach, Germany
Winning Year: 2010
The company's models benefit older workers with care giving needs and offer a variety of courses for older workers, including holistic mnemonic training and retirement transition preparation.

University of Central Lancashire: Lancashire, United Kingdom
Winning Year: 2010
Through its "Fresh Step" program, employees over 50 are encouraged to reflect on their lives through various activities, such as mentoring younger colleagues. Recruitment focuses on age-neutrality and many positions are offered on a job-share basis.

The Co-operative Group: Manchester, United Kingdom
Winning Year: 2011
Measures implemented to remove barriers for older workers include flexible pension options, mandatory training for managers on the benefits of age diversity and anti-age discrimination and recruitment practices aimed at equal opportunities.

Analysis of AARP Initiatives

A deeper analysis of why such model programs are selected by AARP can illuminate program and organizational structures and elements that contribute to work environments that are age-friendly and mutually beneficial to younger and older workers alike.

The structures and elements of workplace programs used in AARP's analysis center on positive workplace policies for mature workers; further, they recognize employees' long-term commitments. Each structure and element highlights the importance of education and training for promoting the employment of older workers. Table 6.1 outlines a number of structures and elements common to AARP-awarded age-friendly programs, and can serve to guide employers in crafting age-friendly workplaces

Another strategy for best practices in managing a multigenerational workforce is the model titled, Make time to discuss; Explore differences; Encourage respect; Take personal responsibility (M.E.E.T) created by VisionPoint Productions Inc. In recognizing that successful organizations excel via productivity, working relationships and customer satisfaction, M.E.E.T capitalizes on common characteristics influencing this success. Therefore, the model focuses on making time to discuss workplace

Table 6.1 Programmatic and organizational structures and key elements of age-friendly work environments

Programmatic + Organizational structures	Key elements
Flexible work and retirement arrangements	Transition from part time to full time; Pre-retirement seminars; Flexible scheduling; Flex-time; Part-time employment; Seasonal work; Consulting; Telecommuting; Job-sharing; Phased retirement; Short-term/special assignments; Restructured salary/benefit packages; Compressed work weeks; Remote working; Mentoring
Lifelong learning opportunities	Openness to continuing success and career development; Education and job-training opportunities; Recognize all forms of learning; Motivate self-directed learning; Ensure all employees have an equal opportunity to excel; Providing honest, rapid, frequent feedback; On-the-job training; Matching skills needs with individual educational status of older employees
Training	Confidence and employability training; Identifying areas at risk for retirement and transferable skills; Foster respect through diversity training (sexism, racism, ageism training); Create an age-diversity task force; Conduct a cultural and demographic audit of your company; Training methods need to take into account changes in information processing with age; Offer development opportunities, such as tuition reimbursement, online training, and certification classes
Recruitment	Focus on potential not age; Utilize placement agencies for mature job seekers; Conduct informational seminars tailored to the older community and places older adults frequent; Form an older worker task force and ask current older employees for referrals and suggestions on how to attract/retain older workers
Promotion of health and diversity	Education on generational differences to create respect in workplace; Ensuring practices are age-neutral; Processes are free from age-discrimination; Addressing demographic changes in the workplace to ensure a well balanced age structure workforce; Smoking cessation programs; Blood pressure testing; Health risk assessments; Fitness classes/health club memberships; Disease management; Weight-reduction programs; Nutrition counseling; Eldercare-oriented benefits; Long-term care insurance; Elder care vouchers; Nursing home care subsidies; Wellness benefits; Accommodations for employees with caregiving responsibilities; Offering unpaid, short-term leave for child and eldercare; Employer-provided elder care services; Corporate retirement housing options

(continued)

Table 6.1 (continued)

Programmatic + Organizational structures	Key elements
Innovation	Encourage and leave room for innovation and allow greater dialogue and engagement among all employees; Find ways to embrace employees' lifestyle realities, rather than just tolerating them.
Consistent with AARP's public policies and values	Commitment to enhancing the quality of life for people as they age in ways that benefit societies as a whole; improve economic security, health care and quality of life
Workplace culture	Build morale; Decrease workplace stress; Improve intergenerational relations; Promoting positive images of aging; Build trust and engagement on the part of the workers; Create an appeals process for grievances (confidential and neutral resource)
Workplace design	Ergonomic work stations or facilities to reduce physical stress on the body; On-site technical assistance; Appropriate workforce literacy; Accommodations including installing ramps and elevators, appropriate lighting; redesigning work spaces, and using technology to assist hearing and vision-impaired workers; handrails; slip-resistant walking surfaces; the use of color contrast between stairway risers and treads
Retiree relations	Creating opportunities for retirees: inviting them to organizational events, recognizing them on anniversaries of their retirement
Work Arrangements and Work-Life Balance	Work schedules; Supervisory relationships; Decision control; Information transfer; Avenues for conflict resolution; Empowerment of the workforce; Autonomy; Delegation of control; Good relations between management and workers; Low stress; Low grievance rates; Encouragement of long term commitment to the workforce
The social context of the work	Greater needs for public transportations, car pools or telecommuting

Chart adapted from AARP (2007, 2010a, b, 2011a, b), Ballone (2007), Ekerdt (2010), Silverstein (2008), and UNECE (2011)

problems, exploring differences, encouraging respect and taking responsibility. This program is designed to help employees understand generational issues in the workplace, work successfully within a multigenerational workplace, minimize the negative impact of generational differences, and maximize opportunities for individual and organizational success (VisionPoint Productions 2006).

The "M.E.E.T. Model" enables employees to recognize how individual and generational differences influence approaches to work, communication, and work/life balance priorities. While learning about these issues and negative stereotypes and assumptions about age, employees are able to recognize how they may affect productivity, teamwork and customer relations.

Because some management styles can be unintentionally offensive to workers from different generations, management and organizational policies must provide a framework for professional behavior that recognizes value differences among workers from different age groups. Additionally, employers must mitigate barriers for job mobility for older workers and work to retain the talents of these workers. This may be accomplished by providing worker and employer information on age discriminations laws; and legislation to discourage early retirement. Employers can also provide information on state-financed training and retraining programs; subsidies for hiring older workers; and information campaigns to build awareness about mature workforce issues (AARP 2007).

Strategies and Policies

Workplace policies and programs can be adjusted to make organizations attractive, safe, and fulfilling for older workers. AARP (2007) has identified eight strategies that can be used by administrators and managers to create age-friendly workplaces. These include:

1. Offer workers maximum work flexibility to respond to the changing needs of the workforce in all stages of the employment lifecycle.
2. Create healthier work environments, ergonomic working conditions, and additional access to health promotion programs.
3. Promote visible leadership related to aging workforce issues.
4. Provide management training that enables leaders to recognize the unique needs of various employee groups while teaching them how to manage and lead a multigenerational workplace.
5. Conduct research related to employees to aid in the understanding of employee behavior, changes in employee perceptions and needs, and the effectiveness of organizational investments and workplace programs.
6. Eliminate age-related stereotypes and empower innovative thinking throughout the workforce.
7. Focus performance requirements on job-related and not age-related characteristics for hiring, promotion and/or retention; and
8. Devote careful attention to strategic talent deployment issues through ensuring equality of opportunity in training and development and providing lifelong learning and growth to employees of all ages to maximize individual and organizational performance.

Additional strategies might incorporate the assessment of ethnicity and values reflected in current organizational practice; survey common assumptions held by managers, supervisors and workers related to age of workers and productivity; analyze workforce socio-demographics and patterns of hiring, firing and leave-taking; survey workers' perception and self-perception of ageist attitudes and treatment; and evaluate programs (if any) in place to address ageism and abuse of workers, including older workers. Examples of strategies include training; human resources department standards; counseling availability; flexible work schedules; improvements in the physical space environment; and intergenerational conflict resolution.

The authors hope that these businesses and models will raise awareness of the reality of the aging workforce as well as effective practices to create supportive work environments that leverage the human capital of older workers (AARP 2007). Although some employers have recognized the importance of effective age management in the workplace, managing an aging and age-diverse workforce remains a growing challenge for many businesses. Employers must recognize the implications of demographic trends and changing views of retirement, and offer flexible and competitive working arrangements and environments to attract, recruit, engage, and retain older workers. In order to create age-friendly working environments, workers must be judged on the basis of skills, competences, motivation and output, and not on age itself. Work environments and legislation need to be revised to accommodate the needs of older workers and eliminate misconceptions about them (Stark 2009). Many employment policies and practices have implicit age biases with regard to recruitment, selection, performance appraisal and assignment to training opportunities. Several authors have highlighted the societal and organizational ambivalence about aging and resultant policy inconsistencies regarding older workers (Gibson et al. 2010; Posthuma and Campion 2009; Stark 2009).

National and local employment policies that are not biased against older workers do exist. Although these policies have the potential to promote age-friendly employment practices and older worker retention, their effectiveness in doing so has not been demonstrated. For example, two recent policy changes have been designed to facilitate the retention of older workers in the workforce: modifications in the U.S. Social Security Program that create incentives for older workers to remain employed; and civil rights legislation to protect older workers from unfair practices and dismissal due to age. To date, policy makers have yet to demonstrate that these policy changes have resulted in retention of older workers. The inability of the Equal Employment Opportunity Commission (EEOC) to bring age discrimination cases to the courts within required time limits and the lack of federally funded job training for the elderly highlight the low priority of this vital issue.

Ekerdt (2010) argued that, with the surge in the elderly population, policies must target sensitive issues regarding the labor market's ability to accommodate, sustain and expand employment of older workers. According to Posthuma and Campion (2009), older employees need to remain in the workforce longer to facilitate national economic growth and the health of the financial system. Therefore, strict policies must be implemented to reduce the ageist stereotypes that inhibit older workers from remaining an active part of the labor force. The economic and social implications of age discrimination in the workplace, retirement and social security are all on the forefront of numerous national debates. In the face of population aging, the demographic and economic projections of older adults must be incorporated into public policy analysis Singer and Manton (1998), and organizational changes to promote supportive work environments.

According to Ekerdt (2010), retirement is a societal practice that primarily manages succession within social institutions, such as families, labor markets and organizations. It should be structured to incorporate the capabilities of the rapidly increasing population of older individuals into our national economy. In addition to the imperative to incorporate these workers into the labor force, the fiscal implications of this growing population have developed into a point of political contention, with a contemporary trend to reduce social programs for old age security (Ekerdt 2010). Paralleling one of the worst economic recessions is the retirement of the baby boomers. Although new policy directions and population demographics encourage the retention of older workers within the workforce, policy makers must also address structural, cultural, and personal forces influencing workplace environments, with the goal of making all workplaces fair and friendly to older workers.

References

AARP (2007). Profit from experience – An international study on the aging workforce. *The Journal*, Premier Edition, 40–41.

AARP (2010a). Best practices from the recipients of the 2009 AARP international innovative employer award. *The Journal*, Summer 2010, 37–46.

AARP (2010b). 2010 AARP-UN briefing series on global aging. *The Journal*, Summer 2010, 99–103.

AARP (2011a). 2010 AARP International Innovative Employer Award recipients. *The Journal*, Winter, 2011, 47–58.

AARP (2011b, September). Best employers for workers over 50 award – international winners. Retrieved December 5, 2011, from http://www.aarp.org/work/on-the-job/info-09-2011/aarp-best-employers-international-winners-2011.html

Administration on Aging (2011). Retrieved November 13, 2011, from http://www.aoa.gov

Bailey, D., Koeny, K. M., McNish, M. E., Powers, R., & Uhly, K. (2008). *Sustaining our spirits: Women leaders thriving for today and tomorrow*. Washington, DC: NASW Press.

Ballone, C. (2007). Consulting your clients to leverage the multi-generational workforce. *Journal of Practical Consulting, 2*(1), 9–15.

Bulter, S. M. (2010). Risking big changes with small reforms. *The New England Journal of Medicine, 362*(8), 673–675.

Bureau of Labor Statistics (2009). The employment projections for 2008–18. Bureau of Labor Statistics, Monthly Labor Review, November 2009.

Chernesky, R. H. (1998). Advancing women into the managerial ranks. In R. L. Edwards, J. A. Yankey, & M. A. Altpeter (Eds.), *Skills for effective management of nonprofit organizations* (pp. 200–218). Washington, DC: NASW Press.

Chernesky, R. H. (2005). Managing in a multicultural environment. In E. Congress & M. Gonzalez (Eds.), *Multicultural perspectives in working with families* (pp. 38–66). New York: Springer Publishing Company.

Clark, R. (1997). *Active aging: A shift in the paradigm*. Washington, DC: U.S. Department of Health and Human Services.

Daily, L. (2011, January/February). My coworker is just a kid! *AARP The Magazine, 16*.

Denison, D. R. (1996). What is the difference between organizational culture and organizational climate? A native's point of view on a decade of paradigm wars. *The Academy of Management Review, 2*(3), 619.

Ekerdt, D. J. (2010). Frontiers of research on work and retirement. *Journal of Gerontology: Social Sciences, 65B*(1), 69–80.

Eyster, L., Johnson, R. W., & Toder, E. (2008). *Current strategies to employ and retain older workers*. Washington, DC: The Urban Institute.

Foggs, S. (2001). *Counting on experience: A review of good practice in the employment of mature worker*. Sydney: NSW Committee on Ageing.

Frost, P. J. (2003). *Toxic emotions at work: How compassionate managers handle pain and conflict*. Boston: Harvard Business School Press.

Garrow, E., & Hasenfeld, Y. (2010). Theoretical approaches to human service organizations. In Y. Hasenfeld (Ed.), *Human services as complex organizations* (pp. 33–57). Thousand Oaks: Sage Publications, Inc.

Gibson, W. J., Jones, P. J., Celia, J., Clark, C., Epstein, A., & Haselberger, J. (2010). Ageism and the baby boomer: Issues, challenges and the TEAM approach. *Contemporary Issues In Education Research, 3*(1), 2–8.

Glisson, C. (2000). Organizational climate and culture. In R. J. Patti (Ed.), *The handbook of social welfare management* (pp. 195–218). Thousand Oaks: Sage Publications, Inc.

Glisson, C., & Hemmelgarn, A. (1998). The effect of organizational climate and interorganizational coordination on the quality and outcomes of children's service systems. *Child Abuse & Neglect, 22*(5), 401–421.

Gustman, A. L., & Steinmeier, T. (2009). How changes in social security affect recent retirement trends. *Research on Aging, 31*(261), 1–31.

Hemmelgarn, A. L., Glisson, C., & James, L. R. (2010). Organizational culture and climate. In Y. Hasenfeld (Ed.), *Human services as complex organizations* (pp. 229–250). Thousand Oaks: Sage Publications, Inc.

Klay, R. J., & Steen, T. P. (2008). Linking retirement and personal responsibility: Social insecurity (pp. 31–37). Chicago: Christian Century.

Laporte, S. B. (1991, January). The sting of the subtle snub. *Working Woman*, pp. 53, 55.

Leonesio, M. V., Bridges, B., Gesumaria, R., & Del Bene, L. (2012). The increasing labor force participation of older workers and its effect on the income of the aged. *Social Security Bulletin, 72*(1), Retrieved from http://ssa.gov/policy/docs/ssb/v72n1/v72n1p59.html

Marshall, V. W., & Altpeter, M. (2005). Cultivating social work leadership in health promotion and aging: Strategies for active aging interventions. *Health and Social Work, 30*(2), 135–145.

Maslach, C. (1998). A multidimensional theory of burnout. In C. Cooper (Ed.), *Theories of organizational stress*. Manchester: Oxford University Press.

Moody, H. R., & Sasser, J. R. (2012). *Aging, concepts and controversies* (2nd ed.). Thousand Oaks: Pine Forge Press.

National Council on the Aging (NCOA). Issues for the 1990s: In a turbulent economy, the U.S. workforce is aging. *Perspectives on Aging*, *20*(2) March/April 1991.

Neumark, D. (2008). *Reassessing the age discrimination in employment act*. Washington, DC: AARP's Public Policy Institute.

Posthuma, R. A., & Campion, M. A. (2009). Age stereotypes in the workplace: Common stereotypes, moderators, and future research directions. *Journal of Management, 35*(1), 158–188.

Rampell, C. (2011a, January 12). More workers complain of bias on the job, a trend linked to widespread layoffs. *The New York Times*, p. B4.

Rampell, C. (2011b, May 29). A generation of slackers? Not so much. *The New York Times*, p. 3.

Rich, M. (2010, September 20). For the unemployed over 50, fears of never working again. *New York Times*, p. A1.

Sanders, S., Dorfman, L., & Ingram, J. (2008). An evaluation of the GeroRich program for infusing social work curriculum with aging content. *Gerontology & Geriatric Education, 28*(4), 22–38.

Schrank, H. T., & Waring, J. M. (1989). Older workers: Ambivalence and interventions. *The Annals of the American Academy of Political and Social Science, 503*, 113–126.

Silverstein, M. (2008). Meeting the challenges of an aging workforce. *American Journal of Industrial Medicine, 51*, 269–280.

Singer, B. H., & Manton, K. G. (1998). The effects of health changes on projections of health service needs for the elderly population of the United States. *Proceedings of the National Academy of Sciences of the United States of America, 95*(26), 15618–15622.

Sparks, K., Faragher, B., & Cooper, C. (2001). Well-being and occupational health in the 21st century workplace. *Journal of Occupational and Organizational Psychology, 74*, 489–509.
Standing, H., & Baume, E. (2001). *Equity, equal opportunities, gender and organizational performance*. Geneva: World Health Organization.
Stark, E. (2009). Lost in a time warp: How age stereotypes impact older baby boomers who still want to work. *People & Strategy, 32*(4), 58–65.
Sutton, R. I. (2010, November 28). How bad apples infect the tree. *New York Times*, p. 8.
Tugend, A. (2010, November 20). Incivility can have costs beyond hurt feelings. *New York Times*, p. B6.
UNECE (2011). *United Nations Economic Commission for Europe Policy Brief on Ageing*, No. 9, January 2011, 1–12.
VisionPoint Productions Inc. (2006). Generations: M.E.E.T. for Respect in the Workplace, http://www.trainingsolutions.com/pdf/Generationspm.pdf
Zunz, S., & Chernesky, R. H. (2000). The workplace as a protective environment: Management strategies. In E. Norman (Ed.), *Resiliency enhancement* (pp. 157–176). New York: Columbia University Press.

Chapter 7
Human Resource Departments and Older Adults in the Workplace

James Woolever

The first decade of the new century has witnessed two noteworthy recessions. Both had a devastating effect on the American workforce.

The 2001 recession resulted in the loss of approximately 2.7 million payroll jobs. Since the beginning of the Great Recession in December 2007, the number of layoffs has reached 8.4 million with only one million workers returning to the workforce. And the situation for American workers is not likely to improve soon. Benjamin Bernanke, Chairman of the Federal Reserve System, does not anticipate the unemployment rate to fall below 6% for another 5 years (Samuelson 2010). Rutgers University professor Joseph Seneca, with Jim Hughes (2009), predicted that it will take more than 7 years to recover the lost jobs following what has become the worst employment setback since the Great Depression. For a full recovery to occur, more than two million jobs must be added every year for seven straight years. This type of economic growth has occurred only twice in this country's history (Seneca and Hughes 2009).

Whenever there is a significant economic recession, pressure is placed on management to cut expenses and downsize the workforce. An obvious target for layoffs and reductions-in-force (RIFs) is older workers who may earn higher wages due to longevity in the job, and might be perceived as risks for higher health care costs. According to government records, approximately 6.9% of workers over age 55 are currently unemployed. Among men 55 years and older, the unemployment rate is more than 8%. On average it takes older workers roughly 33 weeks to find new jobs, nearly 7weeks longer than younger workers, and nearly 13 weeks longer than it took 2 years ago (Close 2009). Additionally, older workers who are reemployed generally do not receive the salary they enjoyed prior to being downsized.

J. Woolever, Ph.D. (✉)
Provost, 1000 El Camino Real, Atherton, CA 94027, USA
e-mail: jwoolever@menlo.edu

P. Brownell and J.J. Kelly (eds.), *Ageism and Mistreatment of Older Workers: Current Reality, Future Solutions*, DOI 10.1007/978-94-007-5521-5_7,

There is no commonly agreed upon definition of "older worker." The Age Discrimination in Employment Act (ADEA) in 2004 uses the age 40 as the demarcation line of an older worker; membership in the American Association of Retired Persons (AARP) begins at 50; the Older Americans Act (1965), the Job Training Partnership Act (1982), and the Workforce Investment Act (2000) use "55 for defining the older worker" (Hedge et al. 2005). A review of 105 studies that employed the term *older worker* revealed that the average chronological age for this particular segment of the population was 53.4 years (Ashbaugh and Fay 1987). For purposes of this chapter, 53 and older defines a worker as *older*. The term *older worker* is now applied to retirees returning to the workforce on a full or part-time basis as well as to individuals making midlife career changes.

The Legal Context

Both federal and state laws prohibit discrimination based on age. The federal prohibition is contained in the ADEA of 1967, enforced by the Equal Employment Opportunity Commission (EEOC). In 1990, Congress passed the Older Workers Benefit Protection Act (OWBPA), which amended ADEA to safeguard older workers' employee benefits from age discrimination. Under this act employers must pay the same amount for each benefit to an older worker as they pay for a younger worker. When dealing with restructuring and layoff programs and early retirement plans, the OWBPA requires employers to provide information about the ages of both terminated and retained employees. (Older Workers Benefit Protection Act, 1990).

Despite well-established laws to prevent age discrimination on the job, a recent Supreme Court ruling has undermined the protection guaranteed by these laws. In *Jack Gross v. FBL Financial Services*, the Court stated that workers need to bear a higher burden of proof to show age discrimination. The Court ruled that the older plaintiff must prove by a "preponderance of the evidence" that age was the decisive reason for the dismissal. For nearly a quarter of a century, Jack Gross was a claims adjuster for FBL Financial in Des Moines, Iowa. He was considered an exemplary employee with performance reviews in the top 3% of the company for 13 consecutive years. In 2003 FBL Financial demoted him and other employees over the age of 50. His job was reassigned to a younger, less experienced, employee. At the time of his demotion, Jack Gross was 55 years old. In April 2004, Gross sued FBL Services alleging he was demoted because of his age. Initially, a federal district court in Iowa found in Gross' favor and a jury awarded him $46, 945, but an appeals court vacated the verdict (Gross v. FBL Financial Services, Inc 2008). Eventually the case made it to the Supreme Court. In 2009, by a narrow majority (5–4) the Roberts' court held that any employee wishing to sue his or her employer for age discrimination must prove beyond a shadow of doubt that age was the deciding factor in management's decision. Prior to this Supreme Court ruling, an employee had only to prove that age was one of the *key* factors for dismissal or demotion. Now the burden of proof lies squarely on the shoulders of the victim to prove that age was the decisive factor in an employment decision.

The Court's controversial 2009 ruling has evoked considerable uproar by members of Congress. Senator Tom Harkin, Chairman of the Senate Health, Education, Labor and Pensions Committee, condemned the ruling claiming that the Court "tore up this decades-old standard and imposed a new standard that the Supreme Court had rejected in a previous case. In its place, it invented a new standard that makes it prohibitively difficult for a victim to prove age discrimination" (Harkin 2010). To remedy this situation, Harkin along with Senator Patrick Leahy of Vermont and Rep. George Miller of California has introduced the Protecting Older Workers Against Discrimination Act (POWADA) designed to restore the previously accepted burden of proof for interpretation of age discrimination in employment. Harkin believes this legislation will "make clear that when an employee shows discrimination was a motivating factor behind a decision, the burden is properly on the employer to show it complied with the law" (Harkin 2010) . POWDA enjoys the full endorsement of AARP and other aging advocates.

Management Practices Targeting Older Workers

The *Gross v FBL Financial* case clearly illustrates the fact that unlawful discrimination in the workplace is often difficult to detect. Although anti-discriminatory laws prevent some targeting for dismissal or demotion on age alone, those employers who discriminate do so in clever ways, often disguising their true intent. For example, various "age-friendly" opportunities promoted by such agencies as the International Labor Organization (ILO) include buy-out packages and other incentivized layoff strategies. Another more popular tactic employed by several large corporations such as Goodyear, Ford, and Capital One is the performance evaluation process known as "forced ranking" (unflatteringly referred to as "rank and yank"). It is estimated that over 20% of U.S. companies, including as many as 25% of the Fortune 500, now employ the forced ranking system. Here employees are ranked against each other based on a particular scheme or design. Normally, the ranking decisions are made by managers who negotiate with each other to satisfy the distribution requirements. At General Electric, employees are put into three categories, with the A group representing the top 20%, B the middle 70%, and C the bottom 10%. Employees in the As are eligible for significant raises and bonuses, the Bs receive more modest raises and the Cs receive no raises. Employees who receive a second C ranking in the next performance appraisal are subject to termination (McCann and Osborne 2004).

During the past decade, a number of Fortune 500 corporations such as Ford, Capital One, and Goodyear have been sued by older workers claiming that the forced ranking system both violates the federal regulations, and is an overt disguise for age discrimination. In a well documented essay, McCann and Osborne concluded that forced ranking intentionally targets older workers for dismissal because they are consistently ranked in the lowest category compared to younger employees. The authors concluded that "by facilitating the actual or constructive discharge of its older workers, the forced ranking system enabled the employer to fulfill its purpose of achieving and maintaining a relatively younger workforce." (p. 8)

Given the recent Supreme Court ruling, older workers now find themselves at greater risk of being abruptly terminated by their employers or to be treated in an unethical manner by management. There is little doubt that the current recession has severely undermined the job security of older workers resulting in emotional and physical trauma to them. With greater frequency companies are "reorganizing" their workforce so that older workers are dismissed, and then told they are ineligible to apply for any new position in the organization. The result is that age-related discrimination allegations by employees are at a record high. According to the Equal Employment Opportunity Commission (EEOC), 24,582 charges were filed in the year ending September 20, 2008, a 29% increase since 2007. EEOC records indicate that overall employment discrimination complaints have risen 15%, with age-based allegations having the highest rate of increase. The number of claims filed in 2009 dropped slightly to 22,778 (Chandler et al. 2010).

Several noteworthy companies have been charged with age-related discrimination. At Lawrence Livermore National Laboratory in California, more than 98 employees are claiming they were dismissed solely due to their age; the 3 M Company has been charged by more than 4,900 employees with age discrimination; and at Dell Inc., four former human resource employees have filed a $500 million age discrimination lawsuit (Lewis 2009). The most bizarre lawsuit involved a 80-year old New York City Home Depot supervisor who was dismissed and replaced with a younger employee over a 24 cent purchase. Ellen Strickland, a 19 year employee, was fired by the company in September 2010, for making a small purchase (four 6-cent screws) in order to get cash from her debit bank account. The problem arose because she failed to take the four screws, and thus the store's inventory was altered. According to management, Strickland's actions "resulted in manipulation of inventory" and she was summarily fired with no chance to be rehired. Although Strickland did not take the four screws and had stellar performance appraisals, she was terminated in fewer than 24 h (Zambito 2011).

A review of the current literature addressing age-related discrimination cases revealed that a number of legal firms are mining a lucrative market offering their services and advice to financial and human resource managers to assist them in avoiding lawsuits by terminated older workers. The focus here is not on assisting managers to adopt better ethical practices toward their employees, but on preventing their companies from being targets for age discrimination lawsuits. These legal firms advised employers who are considering possible layoffs to make sure that younger employees are not transferred in a wholesale fashion to other jobs from which older workers are dismissed. Likewise, employers need to be careful not to offer training opportunities to younger employees while denying the same training to older workers. Managers are reminded that any employee may request job titles and ages of all employees being terminated, as well as titles and ages of employees who are retained.

For employees who take early retirement, legal advisors encourage managers to persuade them to sign a waiver that meets the conditions of OWBPA. In return for signing the waiver, management offers the employees a little extra in their retirement packages than what they are entitled to by right. In this manner, the employee cannot later file an age discrimination lawsuit. Although legal, the ethics of this tactic are questionable. Above all, management is warned that companies lose age discrimination lawsuits when employees can produce evidence that managers or other employees used derogatory language such as "old fuddy-duddy," "obsolete," and "too old to matter" in reference to an older worker. Insensitive comments such as these can be used to support a discrimination claim (Dooley 2004, p. 40). In August 2010, the California Supreme Court in Reid v. Google, Inc. reaffirmed that it was the jury's responsibility to determine what weight should be given to "stray remarks" in an age discrimination claim. This court ruling means that degrading or derogatory language can be admitted into age discrimination lawsuits making it harder for employers to obtain summary dismissals.

Despite a projection by the United Nations that almost one-third of the working-age population in developed countries will be 50 or over by 2050, there is sufficient evidence over the past two decades to confirm that older workers have been disproportionately singled out for mistreatment during organizational downsizing (Armstrong-Stassen and Cattaneo 2010). Older workers are at risk for being stereotyped as being incompetent and disposable based solely on their age.

Organizations often adopt a parochial mindset that views older workers as durable goods with a limited life-span. Employers often perceive older workers as lacking the proper knowledge, technical skills and competencies to fulfill their responsibilities. Additionally, older workers find themselves being unfairly scrutinized in their work-related activities and receiving unacceptable performance appraisals. With employers regarding older employees as a costly liability, companies are unwilling to implement or continue human resources strategies to accommodate the technological and development needs of older workers (Yeatts et al. 2000). Instead, companies either introduce buy-out programs or seek to downsize their older workers entirely.

During an economic downturn there are few incentives to retrain or retain older workers. Employers regard younger workers as more technologically sophisticated and willing to work at lower wages with reduced benefits. The common strategy is to replace or remove older workers rather than attempt to engage them through retraining and development activities (Armstrong-Stassen and Cattaneo 2010). The corporate mindset that depicts older workers as undeserving of training and development opportunities can easily lead to adversarial working conditions and discriminatory practices. When younger employees receive training denied older workers there are sufficient grounds for an age discrimination claim (Maurer and Rafuse 2001). With the average age of the workforce steadily increasing, it is becoming apparent that managers, and particularly human resource managers need to take a different attitude toward treatment of older workers, especially regarding training and development opportunities.

Human Resource Managers' New Role

If older workers are to have their legal rights recognized and if the workplace is to be free of mistreatment or discrimination, the onus is on human resource managers to secure these conditions. Now and in future, managers will have to assume the daunting role of both *compliance* and *ethics officers* in their organizations. It is critical that human resource managers acknowledge and appreciate the difference between these two vital functions.

A compliance officer is principally concerned with guaranteeing that management and employees observe the criminal and regulatory laws applicable to their organization. The task here is instituting safeguards to prevent illegal behavior. In most large organizations the compliance officer is an attorney-at-law who handles reports of misconduct, takes necessary steps to achieve compliance through audits, and enforces regulatory standards through disciplinary measures. The emphasis is on the prevention of illegal behavior by establishing compliance standards and procedures. Compliance officers seek to reduce inappropriate behavior by implementing various training programs geared to educate employees about the imposed norms and standards that govern workplace etiquette. An example is the sexual harassment training that human resource managers oversee so that their companies are in compliance with federal and state statutes. In the past human resource managers generally viewed their role as that of a compliance officer.

No one disputes the importance of compliance training for any organization. At best, today's employees have a rudimentary understanding of the complex employment and regulatory laws that govern their industries. It is the responsibility of human resource managers to educate their employees, especially managers, in current laws governing the hiring, training, compensating and disciplining of employees. While maintaining loyalty to their organization, human resource managers must comply and adhere to all federally and state-mandated laws regarding the treatment of employees.

At the same time laws are often insufficient to guide personal actions and there are a wide range of issues and situations that fall outside the domain of the law. For this reason, ethics plays a role in the business enterprise. Besides upholding all compliance laws and regulations, human resource departments are the "moral compass or conscience" of the organization (Leonard 2011, p. 31). Because the field of human resources is constantly changing and evolving, managers now often assume the responsibilities of an ethics officer in the true sense of the term.

A compliance officer is concerned with what an organization and its employees *must do*; an ethics officer focuses attention on what an organization and its employees *should do.* Legal compliance does not elicit either moral imagination or personal commitment. In itself, the law does not encourage people to commit themselves to virtuous or ethical behavior. The law, per se, seeks to preserve social order and stability; its purpose is not to inspire people to take initiatives based on moral principles. As Lynn Sharpe Paine, a Harvard University ethicist argued, "Those managers who define ethics as legal compliance are implicitly endorsing a code of moral mediocrity for their organizations" (Paine 1994, p. 111). Organizations must move away from an "avoidance of legal sanctions" mindset to one that endorses and demands ethically sound behavior throughout the company. For this to occur, human resource managers need to be proactive in creating and implementing an organizational culture that prizes moral integrity. A new era is dawning where human resource managers will have to assume the role of ethics officer in their organizations in order to guarantee that employees are treated in a just and equitable manner.

For a company to evolve from being a legalistic to an ethical organization, both upper management and human resource departments need to demonstrate an active and consistent concern for ethical policies throughout their operating systems. This is especially true in personnel matters. Decisions need to be made based not solely on anticipated economics outcomes, but on what is the "right thing" to do in each situation. Here ethical values become more than fanciful words in a mission statement; rather, they serve as the core values for guidance when dealing with complex internal and external issues. Ethical analysis becomes an integrated partner in developing the company's vision and strategic plans. All phases of organizational policies and procedures must reflect the stated core ethical values (Reidenbach and Robin 1991).

There is no infallible method that human resource departments can employ to create an effective integrity strategy within their organization. Yet there are key markers that epitomize committed ethical organizations. Lynn Paine identified five hallmarks that typify an organization committed to moral integrity:

- *Guiding values and commitments are clearly communicated.* These values explicitly state the organization's commitments to its various stakeholders and appeal to its members. Employees take shared values seriously and feel free to discuss them in the workplace.
- *Company leaders are personally committed to acting on the core values.* Leaders consistently utilized the core values in making policy and determining priorities. Furthermore, leaders do not waffle when accepting responsibility for making difficult ethical decisions.
- *The values are incorporated into the routine-decision making process and are reflected into every important organizational activity.* Ethical core values help shape all phases of human resource activity, including goal setting, allocation of resources, gathering and dispersal of communication, evaluation of performances and job promotion.
- *Systems and structures support and reinforce the company's core values.* Information systems are designed to serve the organization's values, and to guarantee that accurate information is conveyed regarding performance appraisals.
- *Managers have the necessary knowledge and skills to make ethical decisions on a daily basis.* In order to make difficult decisions, organizational leaders, including human resources managers are provided with ethics training and education as a integral part of their professional development (Paine 1994 p. 112).

Over the past two decades, several companies have attempted to implement various integrity strategies with varying degrees of success. Notable companies such as General Dynamics, Martin Marietta, Southwest Airlines, Starbucks, and

Hewlett-Packard have initiated integrity-based ethics programs. Moving a company towards a strong commitment to corporate integrity has both short and long-term benefits. There is a direct connection between corporate ethics and profitability.

According to Howard Putnam, former CEO (2009, p. 3) of Southwest Airlines, "Your ethics and integrity will determine your reputation and your legacy. They are your greatest assets. Don't let anybody ever tell you differently." When senior managers assign a monetary amount to their company's public reputation they realize that the long-term financial costs of doing business unethically are enormous. Losing one's reputation means losing business. The CEO of Texas Instruments, Jerry Junkins maintained that a company's reputation is a priceless asset. This is especially the case for treatment of employees, in particular older workers. When employees witness dishonesty or mistreatment by managers the consequence is a "rotting of the organization." "There is no way," Junkins states "that you're going to be able to rebuild credibility with those employees when you're trying to energize an organization to go do something else. You've created a permanent problem in terms of how people view you as an individual and how they view the management of the organization" (White, n.d., p. 3).

Ethics Committees and Codes

The implementation of moral integrity into an organization demands that concrete and tangible measures be put into place by senior management. Depending on the size of the organization the creation of an ethics committee to oversee compliance and ethical issues sends a clear message that ethics is an integral part of the organization's daily operations. It is the collective responsibility of the ethics committee to develop and implement ethics programs and a code of ethics that will be adopted by the Board of Directors. The aim of the committee is to ensure that both regulatory and ethical standards are established within the organization and that they are communicated to all stakeholders. The ethics committee should also investigate any moral dilemmas or allegations of misconduct by management or employees. For this reason it is imperative that members of an ethics committee be representative of the entire organization. The committee members should include senior officers, human resource managers, and employees. Committee membership must include all age groups. Human resource managers should encourage older workers to serve on ethics committees. It is critical that older workers feel a sense of participation in important ethical decision-making processes. Having a committee to address difficult ethical decisions, especially in matters associated with employee grievances, tends to produce just, equitable decisions if diverse perspectives and age groups are included in the committee. Moreover, inclusion increases the credibility of the decision making process by reducing suspicion of prejudicial bias. To prevent and avoid age-discrimination practices or age-related stereotyping, older workers need a point person, such as a member of the ethics committee, whom they can approach with their ethical concerns.

One of the first tasks of an ethics committee is to design a code of ethics that clearly articulates business principles and standards of conduct that are specifically tied to the organization's mission and objectives. A code of ethics establishes goals and standards, and identifies the core values that the organization commits itself to uphold when conducting business with all its internal and external stakeholders (Sullivan 2009). In formulating a code, the ethics committee needs to solicit views from all stakeholders so the final version articulates ethical issues and concerns that are likely to arise in the foreseeable future. It is also the ethics committee's responsibility to approve the final version of the code and to make sure that it is properly disseminated to all interested parties. Additionally, this should be the group to clarify and interpret the code, facilitate its use, investigate grievances and violations of the code, and discipline violators.

The development of a well stated code of ethics is a time-consuming, pains-taking endeavor. Yet the finished product is worth the effort. On the one hand, it becomes the guiding beacon for responsible business practices; on the other, it identifies specific procedures to be utilized when confronted by ethical problems. Most importantly, a code provides a human resources department with a practical hands-on yardstick for dealing with a pyramid of work-related issues such as age, race or gender-discrimination grievances, conflict of interest practices, employee-management relationships and performance reviews (Sullivan 2009). Publishing a code of ethics is a sound tactic to prevent personnel mistreatment in the workplace. In the post-Enron era the majority of mid-sized and large American companies have adopted some form of a code of ethics.

A survey of codes of ethics revealed that they fall into two categories: *inspirational* and *prescriptive* (Sullivan 2009). Inspirational codes seek to articulate virtues that are prized by stakeholders, such as trustworthiness, respect, compassion, justice and courage. The limitation of this type of code is that it can lack specificity and the language can be vague and subject to individual interpretation. Prescriptive codes offer more specificity by identifying key behaviors that must be adhered to by management and employees alike. The wording is direct, straightforward and indicative of key compliance regulations required in a particular industry. Although a code of conduct cannot prescribe preferred behavior for every problem that may arise in the workplace, it can address concrete issues such as confidentiality, drug use, age-discrimination, conflicts of interest, sexual harassment, whistleblowing, and respect for co-workers. Because a code serves as an "ethical document," it should go beyond compliance expectations and identify the core values that personify the organizational culture. In order for it to be practical, it should include wording that indicates where employees can go if they have questions. Once the code has been approved by the Board of Directors, it should be distributed to all employees, posted on the organization's internet site, and extensively utilized by the human resource department in all training activities. Codes are a document-in-development: thus, they need to be reviewed periodically and updated on a yearly basis. Code of ethics should not be viewed as a panacea that will prevent all forms of unethical conduct or personal indiscretions; rather, they are platforms for an organization to articulate its commitment to responsible business activity.

Human Resource Policies & Procedures

After the code has been announced and distributed, it is the responsibility of human resource departments to update personnel policies and procedures regularly to produce the preferred behavior outlined in the code of ethics. To ensure conformity to the code, personnel policies and procedures should include job descriptions, performance appraisal forms, management-by-objectives expectations (including expectations about participation in training programs), budget report formats, and other key indicators that ensure conformity to the code of ethics.

Managers need to articulate expectations about ongoing employee training in areas of new technology. To prevent mistreatment or discrimination of older workers, policies and procedures need to specify the amount of ongoing training expected of all employees. Older workers have a right to know in their job description what is expected of them concerning advanced technical knowledge and skills. A common grievance of older workers is that they are not afforded sufficient opportunities for training in new technology or not informed that familiarity and adaptability of advanced technology skills are an important criteria in their performance appraisals. Human resource departments need to define and inform employees clearly of the consequences for not utilizing the opportunities provided for ongoing training.

Ethics training provides an important distinction between *invincible* and *vincible* ignorance in terms of moral culpability. For example, if employees are not made aware by management or human resource departments that ongoing training and education are a necessary condition for retention, they have legitimate grounds to complain if they receive a poor performance appraisal due to non-participation in training or failure to update skills. On the other hand, if the job description clearly identifies ongoing training as an integral part of their job duties, employees, in particular older workers, must bear the full responsibility for failure not to avail themselves of these opportunities.

As codes of ethics become integrated into the business culture, human resource managers must be active in providing older workers with incentives and opportunities so they feel committed to the goals of the organization. Policies and procedures need to be put into place that provide support and encouragement for older workers' work performance, intrinsic motivation, and physical and psychological health (Hedge 2008).

Human resources departments should be vigilant to combat age stereotyping. As evidenced in the *Reid v. Google* lawsuit, there remains a proclivity among young workers to make inappropriate comments towards older workers. Brian Reid, a former director of engineering, worked at Google from 2002 to 2004. He left his teaching position at Stanford University to work at Google when he was 52 years old; he was well qualified with a doctoral degree in computer science. During his brief tenure at Google, Reid was told by a 38-year old supervisor that his opinions were "obsolete" and "too old to matter." Reid's work conduct was criticized as "slow," "sluggish" and "lethargic." To worsen matters, Reid's co-workers referred to him as an "old man," "old guy," and "old fuddy-duddy." One of his co-workers remarked that Reid's office placard, which was a picture of a compact disc, should be an LP (Reid vs. Google 2010).

It is this type of adolescent behavior that cannot be tolerated by a company committed to organizational integrity. This stereotyping provides a foundation for age discrimination bias and mistreatment of older workers. An important aspect of managerial training must be to identify myths and inaccurate generalizations regarding the aging process and the capabilities of older workers. Managers need to be trained to recognize the signs and symptoms of age-stereotyping and prevent it before this prejudicial bias metastasizes throughout an organization.

Studies indicate that age-stereotyping appears in two forms. The first and most obvious is *"age stereotyping"* referring to the misconceptions people have about age and worker capabilities. Here older workers are routinely denied access to training or development activities because managers assume they are either unable to learn new forms of technology or not interested because they soon will be retiring. The second misguided bias is *"age norming of jobs,"* which involves establishing a correlation between job duties and the "typical" age of the worker performing those duties. Here some jobs are viewed as oriented for younger workers and other as suited for older workers (Hedge et al. 2005). The danger with age norming is that it falsely assumes that some jobs are the "domain" of particular age groups. It can easily lead to prejudicial bias that some work is beyond the ability of workers solely based on their age. Both types of stereotyping can result in flawed work appraisals that are based more on age than on performance.

In order to prevent these biases, human resource departments need to perform "age audits" regularly to determine the distribution of employee ages for the various positions within the organization. There also should be ample distribution of older workers throughout the organization, including human resource departments. Finally, older workers should be actively engaged in the ethics committee to address and combat any negative stereotyping directed toward their peers (Hedge et al. 2005).

Human Resource Strategies for Older Workers

For better or worse, the American workforce is aging. A large percentage of the workforce is approaching the traditional retirement age of 65; and the number of younger workers is decreasing (The Aging Workforce in American, n.d.). At the present time, workers who are 40 years old or older represent 48% of the total U.S. workforce. Demographers project that between 2004 and 2020, the number workers age 55 and over will increase by 80% to over 33 million people (Technology Trends 2003). In 2001, the number of workers age 40 and above surpassed the number of workers 40 and below for the first time. According to the Labor Department's household survey in December 2010, 28.2 million workers over 55 had jobs, an increase of 7.6% from 3 years.

Thirty years ago, one in seven U.S. jobs was occupied by a person 55 or older. Today the proportion is one in five (Norris 2011). There are several reasons for this change. First, people are living longer and working past the traditional retirement age. Second and most importantly, fewer older workers are willing to retire, despite the high unemployment rate among younger workers. An AARP survey revealed that 69% of employees over 45 had no intention of retiring at 65 (The Aging Workforce in America, n.d.).

Business leaders, managers, and policy makers increasingly agree that companies need to retain and hire older workers. According to Hans-Paul Burkner, CEO of Boston Consulting Group, at the 2010 World Economic Forum in Davos, "We have to give people a chance in their 60s and early 70s to continue to work. We have to have different opportunities for them to expand their talents and to make the best use of their talents" (Roth 2011). Burkner further argued that you need to survey your present workforce and determine where the knowledge worker gaps are going to be in a few years. Managers and human resource departments should monitor closely the age profile of their workforce to determine how impending retirements will affect their companies. When older workers retire or are fired they take with them experience that is irreplaceable. A 2005 Towers Perrin study sponsored by AARP found that the probable benefits of a stable workforce, with older workers retained, combined with reduced turnover costs – due to recruitment, training, onboarding, and getting people up to task – often exceeded the incremental compensation and benefits cost normally associated with keeping older workers (Byham 2008). Burkner urged hiring managers to think carefully about hiring cheap, young labor at the expense of workers who have a lifetime's worth of skills (Roth 2011).

Burkner's challenge to keep older workers in the workplace has immense implications for human resources departments. One result of the aging of the population is that the number of managerial leadership openings will outstrip the available talent. Businesses will need older workers not only because of the impending shortage of younger workers to replace them, but also because of the need for the skills and expertise of seasoned workers. Employers view older workers as stable, honest, dependable, loyal, with a strong work ethic toward their jobs, and less likely to miss work or abruptly leave the workforce. The value of the older worker was confirmed by Pew Research Center (2010) survey in which nearly six-in-ten respondents (employers, managers and human resource personnel) cited work ethic as one of the big differences between young and old and with about three-fourths of the respondents indicating they believed that older workers have a better work ethic. Management is slowly acknowledging that mature and older workers are more efficient in their job performance, make fewer mistakes than younger employees, and despite age-stereotyping, demonstrate that their job skills do not diminish with age.

Penner et al. (2002) found that older workers do not stay in the workforce strictly because of pay and benefits. Instead, they remain employed because they enjoy the internal rewards of work, such as the feeling of accomplishment, of learning new skills, and participating in an important project. Older workers have much to offer organizations. The popular negative myths and stereotypes of aging workers threaten productivity and morale, and contribute to the underutilization of this employee population (Kaye and Cohen 2008). Successful twenty-first century companies cannot afford to squander personnel resources if they wish to compete in the global marketplace.

To engage older workers and maximize their performance, human resource managers need to be flexible in their assignment of workers to tasks. There are a number of concrete strategies that are available to managers, but they must begin by "looking at their enterprises and spotting creative uses for their current (older) workers" (Roth 2011). Instead of keeping older workers in jobs that are stale and routine, they need to be shifted into new positions with responsibility and value. Of all the things managers can do to stimulate the creativity of older workers, the most efficacious is matching personnel with the right assignments – a process of personnel alignment. Human resource departments can provide an invaluable service by matching older workers with jobs that correspond to their expertise and creative thinking skills and that ignite their intrinsic motivation to be more innovative and productive in their jobs. Corporate creativity is more dependent on proper job alignment than any other aspect of business or management. All organizations - profit or nonprofit – have to be strongly aligned in order to be consistently innovative (Robinson and Stern 1998). Alignment of older workers is both intangible and hard to achieve. It is more of a rarity than the norm in most organizations. It can only occur when human resource managers possess accurate information about each of their employees and their proven skills to determine if they have been given the right assignment. Gathering of this information is difficult and time consuming, but well-worth the effort if it results in alignment of workers with appropriate jobs.

Flexible Workplace Opportunities

If human resources departments seek to attract and retain talented older workers they will need to develop concrete strategies to keep them motivated and productive. A vital element of such strategies is meeting the desire of older workers for flexible work options and schedules. A survey conducted by the Sloan Center on Aging & Work found that over 78% of the respondents reported that having access to flexible work options contributes to their satisfaction as employees to a "moderate" or "great extent," and 90% reported that flexible work options contributed to their quality of life to a "moderate" or a "great extent" (Workplace Flexibility 2010). For older workers, the traditional nine-to-five, 5 day work week is no long their preference (Pitt-Catsouphes and Smyer 2005). The same study concluded that flexible schedules did not lower the effectiveness of employees. Employees who enjoyed flexible schedules received favorable appraisals by their supervisors. A flexible workplace might offer employees flexibility regarding the following: work hours, compressed work weeks, work location, job sharing, part-time work or phased retirement, and redesigned work assignments.

Flextime

Flextime is a system of flexible working hours and can include a number of options. The goal is to allow employees to vary their working hours in order to create a better balance between work responsibility and personal commitments, such as caring for a sick family member, attending a grandchild's soccer match or volunteering for a community activity.

There are multiple options now available to create flexible work schedules, including changing the starting and quitting times, and allowing the worker to work at specified "core" hours with the other hours flex-scheduled by the employee. In the latter case, older workers might prefer to work their "core hours" early in the day or in the late afternoon. Employers can also give employees a choice of shifts, control over break-time, or permission to work a compressed workweek for at least part of the year. These are attractive options for workers during the summertime. Flextime scheduling has become a popular and widely utilized option for both employers and employees. A Ranstad survey (2008) found that flextime scheduling was regarded as one of the top three benefits employees identified as promoting personal happiness and fulfillment in the workplace. Flextime scheduling is an effective tactic for lowering stress, reducing absenteeism and tardiness, and improving job satisfaction (Hedge et al. 2005).

Flexlocation

Employees use *flexlocation* to allow employees to fulfill their job responsibilities away from the workplace - at home or at a satellite location on a full-time or part-time basis. Telecommuting allows older works to conduct their work off-site via telecommunications. The option of flexlocation is now utilized by about one-third of all employers (Bond et al. 2005). Both employees and employers view this as a win-win alternative to working in the traditional workplace. For employers, it decreases costs for office space, and for employees it reduces the stress of commuting to work on a regular basis. Older workers have indicated that a flexlocation option is likely to delay their retirement. (Parkinson 2002). This is also an attractive option for older workers who have care-giving responsibilities or are in a phrased retirement.

Job Sharing, Part-Time Work or Phased Retirement

Having two workers split one full-time job allows employees to share their unique talents, abilities and perspectives with each other, and to split the compensation and benefits for that position. *Job splitting* between older and younger workers provides older workers with the opportunity to coach and mentor younger employees. In this manner older workers are recognized for their experience and wisdom. As a result, they are more likely to remain in their positions and their contributions counteract negative stereotyping by other workers and by managers and employers. Another benefit of job splitting is that the well-being of older workers can be increased by giving them an opportunity to work fewer hours. By implementing flexible job hours or allowing them to work on a reduced schedule, employers have a better chance of keeping older workers.

Part-time work is typically classified as work that demands 35 h per work or fewer with limited or no benefits. For those older workers who do not require a full-time position, part-time employment is a viable option for those who wish to scale back their work obligations and begin a *phased retirement* strategy. Phased retirement is a form of part-time work designed specifically for older workers. A phased retirement strategy allows older workers to reduce their work hours or work at a lower classification level without changing companies, and is a useful transition into retirement. The principal advantages of phased retirement programs for employers is the retention of well-trained personnel, reduced costs for training new employees, and lower salary and benefits for older workers shifting from full-time to part time status. The advantages of phased retirement programs for employees are flexible work schedules, a gradual transition into retirement, and the opportunity to supplement retirement income (Brainard 2002).

Although studies indicate that older workers (aged 55–74) prefer to work fewer hours (Moen et al. 2000), only 14% of American companies have a formal phased retirement policy (Hutchens 2003). Unfortunately, pension rules continue to prevent older workers from having phased retirement programs at their workplaces. These regulations make it almost impossible for employees to maintain their previous income using a combination of social security, pension, and paychecks. Furthermore, strict pension rules prohibit companies from giving partial payment to employees who seek to reduce their work hours before they reach retirement age.

Given these restrictions, formal phased retirement programs have been a rarity in most industries. Companies offering phased retirement programs are usually high-tech firms which have a critical need for highly skilled workers. An example is a Silicon Valley company, Ultratech, Inc., which makes photolithography systems. In 2009, Ultratech instituted a phased retirement plan to prevent the growing loss of highly qualified older workers. Under this plan, employees who are 50 and older have the option of reducing their work schedule or working as contractors. To retain their older workers Ultratech's human resources department implemented a "truck system approach" for training younger employees – that is, an older worker has to "get hit by a truck" before someone younger is trained for his or her job. The purpose of this unique strategy is to force Ultratech's managers to create a transition plan for retiring employees and for mentoring younger workers. The goal is to keep new retirees actively involved in the company, and to feel appreciated (Gale 2011).

Job Training

The retention and productivity of older workers in today's workplace will ultimately depend on the quality of the training and education they receive from their employers. It is critical that older workers be allowed to participate in training programs on a regular basis. Education and training are an invaluable resource for guaranteeing the success of any twenty-first century company.

Whenever Jack Welsh, retired CEO of General Electric, is queried on how he was able to able to develop such a dynamic and energized workforce at GE, his answer is always the same: Crotonville! Established in 1956, this corporate training facility in upstate New York has become a career defining event for GE's entry-level and upper management employees. According to Warren Buffett, the Crotonville training facility is the principal reason General Electric has developed top-flight managers and employees for the past 50 years. The mantra at Welch Leadership Center in Crotonville is: *never stop learning.* Its educational and training programs are designed with a twofold purpose: encourage exploration and connect with your fellow learners (Leadership and Learning, n.d.). The Crotonville model is one that human resource managers need to examine as they attempt to create innovative training and retraining programs for their employees, in particular older workers.

Despite the obvious benefits of ongoing education, surveys indicate that compared to younger workers, older workers receive a smaller share of both on-the-job training and contract education (Zeytinoglu et al. 2007). Furthermore, training opportunities are less frequent for the older employee workforce. For example, government statistics verify that that 55- to 64- year-olds were likely to receive one-third of the training offered to 35–44 year-olds (Maurer and Rafuse 2001). Older workers tend to be excluded from training because managers assumed that they cannot learn, are unlikely to learn, or will soon be retired thus reducing the importance of training. One of the classic stereotypes leveled at older workers is that they lack the intellectual dexterity and enthusiasm to master new technologies. This bias was evident in a survey of 400 human resource managers, where only 17% of the respondents rated older workers' learning skills as either "excellent" or "very good" (Office of Technology Assessment 1990). The perception that older workers are uncomfortable with new technologies has been a widely accepted truism for the past 20 years.

To a degree, older adults have contributed to this stereotyping by their lack of interest and resistance to job training. At the same time, both the lack of training opportunities and lack of support by upper management has created an age-unfriendly environment for many older workers. In the past, flawed research studies indicated that older workers were either incapable of learning at the same pace of younger workers or were less interested in learning new skills. When older workers are consistently portrayed as lacking the intellectual prowess to improve their technological skills, the result is a negative psychological effect on their attitude toward training. This age-stereotyping leads to a loss of self-confidence and a diminished desire to pursue training opportunities (Maurer and Rafuse 2001).

Older workers need to be reassured that they are more than capable of learning new skills, especially as related to technology. Fortunately, age-specific studies indicate that older workers are capable of acquiring new skills in computer technology, but the process may require take longer and require hands-on training (Office of Technology Assessment 1990). Older adults learn at a different speed than younger workers. They require more time to acquire the information or skill and to demonstrate that learning has occurred. Pedagogical approaches that accentuate self-paced learning with ample opportunities for practice have proven to be successful for mature adult learners (Yeatts, Folts & Knapp). The physical and mental changes that do occur in workers can be compensated for by effective training designs. When human resource departments select or design training programs they need to be responsive to age-related physical and mental changes among workers.

There are several strategies that human resource managers can employ to ensure success for educating and training an aging workforce:

- Encourage older workers to participate in training and learning activities, including job rotation, coaching, and all related development opportunities. Frequently discuss the importance of ongoing training, including new technologies and procedures that affect performance in the workplace.
- Hold older workers to the same training standards and evaluation instruments as young employees. Document attendance. Provide continual feedback, including progress reports and practical tips for improvement. Keep setting performance goals for older workers and do not lower expectations because of age. Require passing scores at the end of the training, and publicly recognize workers' learning achievements.
- Conduct training sessions in a comfortable environment and make sure that there is adequate lighting and good acoustics. Visual aids should have easy-to-read print with high contrast. At the beginning of each session the trainer should check to be sure that everyone can hear the presentation. The trainer should use a microphone if necessary.
- Keep class sizes small so that more personalized attention can be provided by the trainer. This is particularly important when teaching technical skills. Additionally, shorter sessions are more effective for all workers. Bear in mind that the brain can only absorb what the seat of the pants can endure. Training sessions should allow participants to take breaks on a regular basis.
- Keep the training process simple and straightforward, emphasizing concrete concepts over abstract ideas. Provide real-life examples and practical experiences of participants to illuminate key points. Encourage participants to contribute to various learning exercises with their own experiences and to disagree with concepts that are contrary to their own on-the-job practices.
- As a way to reduce anxiety among older workers, conduct computer-based training well in advance of when the new skills will be needed. Because older workers need more time to assimilate new skills, favor practice-based learning. There should be ample opportunities for practice during the learning sessions. In this manner participants gain confidence in their ability to master new technologies.
- Allow older workers to master the material and exercises in an open, self-paced environment. Because adult learners are self directed, be careful to allow older learners the freedom of self-discovery and the need for professional guidance. There is a fine line between needed guidance and undesired direction. Mature learners enjoy a task-oriented environment and learn by doing and also by failing. Life experiences have taught them that failure is often the necessary ingredient

for self-improvement. For this reason trainers need to demonstrate patience and allow participants to pace themselves, even if they tend to be slow at times.
- Ensure that the self-esteem of all participants is respected at all times. Adult learners are sensitive about their ability to absorb new material. They do not like to be shown up as slow or incompetent. They need continuous reinforcement for their positive steps towards mastery of new technologies (Klatt 1999).

Today's workers recognize that training and ongoing education are critical for their continued tenure in their jobs. AARP surveyed 400 business executives on what they regarded as the three key issues confronting today's workforce. The three most commonly mentioned were: "Training your current workforce," "Introducing new information technology into the work environment", and "Retraining your workforce" (AARP 2000). Successful work-based learning and training opportunities for older workers have the potential to improve employees' motivation for greater self-confidence, and ultimately, greater productivity in their jobs. It is the responsibility of human resource managers to ensure that older workers are given full and equal access to training programs and that these programs are tailored to their unique characteristics.

Providing learning opportunities for all workers is a legal and ethical necessity for human resource departments. Companies seeking to avoid age discrimination charges need to be mindful of the Todd Maurer and Nancy Rafuse (2001) warning that "if older workers are denied training while younger workers receive training, then age discrimination may be inferred" (p. 113). Instead of allowing the job skills of older workers to deteriorate and become obsolete, enterprises need to create a supportive climate for ongoing training and development. Human resource departments have a moral obligation to counter any organizational mindset that views mature and aging workers as being incapable of retraining or not meriting the investment. Managers can prevent age stereotyping and combat obsolescence by removing the barriers to training programs and positively promoting the benefits of continuous learning by providing tangible incentives for older workers, such as reimbursement of tuition costs and positive appraisals. Like their younger counterparts, older workers want promotions leading to higher wages (Helson and Cate 2007). Employees need to be rewarded not only for achieving performance goals, but also for the continuous acquisition and development of new technical skills (Hedge et al. 2005). Yeatts, Folts & Knapp (2000) correctly noted that age alone cannot account for obsolescence of knowledge or skills. In most instances the real culprit is the lack of foresight on the part of employers to deliver state-of-the-art training programs to their workforce. Older workers accept that they must be life-long learners. They need the opportunity to prove to management that they are just as capable of learning and adapting to new technology as younger workers. Human resource departments should collect relevant data about specific job expectations and then identify the ongoing educational and training needs required for these positions. In this manner education and training can be incorporated into yearly performance appraisals. This is an effective method to monitor and avoid skill and knowledge obsolescence.

Older Worker Engagement

In recent years the issue of employee engagement has been the subject of a great attention both in the workforce and in academic research centers (James et al. 2007). Business leaders are keenly interested in employee engagement for its impact on company profitability. The lack of engagement by American workers is costing American businesses approximately $300 billion a year in lost productivity (Saks 2006). Hence, it has become a top priority for executives to identify the cognitive and emotional factors that motivate employees to maximize their creative talents in the workplace. It is difficult enough to align employees to the right job setting; it is a herculean task to motivate workers to perform to their full potential. Transforming individual human talent into collective innovative productivity begins with putting in place appropriate systems and procedures that identify creative efforts as a top priority.

Although there are various definitions of *"engagement,"* it normally refers to the cognitive, emotional and behavioral dispositions associated with an individual's performance in the workplace (Kahn 1990). Engaged employees view themselves has having the intellectual expertise to deal with changing demands of the job, are emotionally invested in the success of the organization, and are willing to put in extra time and effort needed to help the organization accomplish its goals (James et al. 2007). Engaged workers are "passionate" about their work activities. They are energized workers who seek to energize those around them. They take pride in themselves as professionals and take pride in their work.

If the Labor Department's grim projections about the impending labor market shortage come true, many American companies in the next 10–15 years will lack key management and administrative talent which will have serious consequences on their productivity and profits. Although organizations may continue to harbor negative ageism stereotypes, they need older workers to survive (Dychtwald et al. 2004). Instead of encouraging retirement, they need to discourage it. According to Glenn McEvoy and Mary Jo Blahna (2001), the challenge facing human resource departments vis-à-vis older workers is threefold: "retool," "retain," and "reengage." They argue that if older workers are successfully revitalized and retooled for new or expanded assignments, companies will more likely want to retain them. The Sloan Center on Aging and Work at Boston College has conducted extensive studies on employee engagement and discovered that the "primary driver for predicting employee engagement, for both professional and hourly older workers is supervisor effectiveness" (James et al. 2007, p. 4). This conclusion is reaffirmed by the McEvoy & Blahna study that focused on the problem of disengagement by mature workers. Here again the root cause for disengagement was a perceived lack of respect from supervisors and disagreement with upper management (McEvoy and Blahna 2001, p. 48). Supervisors who were supportive, respectful, and demonstrated personal concern for older employees motivated them to higher levels of performance and delayed retirement considerations (James et al. 2007, p. 5). On the other hand, when supervisors, especially those younger in age, disregarded the opinions of older workers, failed to solicit their input on critical issues, and reassigned them to dead-end jobs, aging workers tended to view retirement as the best alternative (McEvoy and Blahna 2001, p. 48). Managers need to realize that after financial income, the second most important reason for older worker to stay in the workforce is emotional fulfillment (Crampton et al. 1997).

In a study of older workers, James et al. (2007, p. 7) reported that when judging the effectiveness of their supervisor, older professional workers (55+) identified five primary drivers of employee engagement:

Opportunity for promotion and career development
Schedule input
Flexibility
Perceived fairness
Job alignment

For older hourly workers, supervisory effectiveness was determined by

Schedule input
Flexibility
Job alignment
Opportunity for development

For professional workers it is important that promotion opportunities are distributed in an equitable fashion. Otherwise, both professional and hourly workers identify the same four drivers as seminal factors in determining whether a supervisor is effective. Despite the long-standing stereotype that older workers are less interested in training and acquisition of new skills, the Sloan Center's extensive study on employee engagement for older workers disproved this myth. In this study older workers welcomed the opportunity to learn new technology with as much interest and enthusiasm as younger employees. The difference is that older workers were not always provided with equal access to these education and development opportunities. Fully engaged older workers are the ones who have a competitive chance for ongoing development and promotion within the organization.

Studies conducted by the Sloan Center on Aging and Work have shown that older workers, professional and hourly, no longer desire the typical 40-h workweek. This is a critical factor in their decision to remain in the workforce and their level of engagement. When organizations make the connection between workplace flexibility and overall effectiveness, they are likely to provide employees with flexible work options. Providing flexible work options is an effective strategy to recruit, retain and increase job engagement of older workers.

When older workers wish to remain in the workforce past the traditional retirement age, their preferences about work schedules need to be considered by their supervisors. Surveys show that the "responsive" supervisor of older workers is the manager who demonstrates flexibility regarding scheduling issues and is supportive of flexible work options. (James et al. 2007). Additionally, the "responsive" supervisor is one who provides support in the face of work-related problems, offers recognition and praise for good work, demonstrates sensitivity about the effect that work schedules have on personal and family life, encourages innovation, and clearly identifies criteria for performance appraisals (p. 7). The bottom line: engagement and retention of the older worker is dependent on managerial effective and workplace flexibility.

Conclusion

A review of the literature on the aging workforce for the past 20 years reveals a number of consistent themes that need to be addressed by human resource departments if they wish to discourage retirement and reengage older workers. Organizations will need to change in order to counter prevailing negative stereotypes about this age group. This is an issue that an ethics committee should address in a timely manner. Supervisors and managers will also require training in how to maximize the long-standing skills and work-related experience of older workers. The degree of respect and deference shown towards older workers by their employers and supervisors is a critical factor in older workers' decision to take early retirement. Many of the recommendations set forth by research centers to keep older workers engaged and committed are the same criteria offered by management experts to enhance creativity and innovation in for-profit and non-profit organizations (Amabile 1999). The key drivers of employee engagement and commitment:

- Alignment: Match employees with the right assignment that plays to their expertise and creative thinking skills. Make sure that work assignments are challenging and important and that older workers know their work makes a difference.
- Freedom: Allow older workers the freedom to determine how best to complete their projects. Allow them to approach problems in ways that will stretch their abilities.
- Training: Give older workers the same opportunities as younger workers for training and learning development in new technologies. Design training sessions which will allow older workers to gain new technological skills in a self-paced manner.
- Supervision: Provide older workers with supervision that is supportive, enabling, and sensitive. Display a sensitivity to assist older workers who are dealing with difficult personal or family issues.
- Supervisory Encouragement: Generously recognize the unique efforts of older workers. Encourage older workers to mentor new and younger workers by passing on their accumulated experience and wisdom. Given their experience and work ethos, older workers are well suited to serve as role models.
- Organizational Support: Implement appropriate systems and procedures to create a work environment that is both challenging and enjoyable. For job satisfaction late career workers need flexible work options. Finally, communicate regularly to older workers that their contribution are critical for the success of the organization, and provide them with incentives to remain in the workforce (McEvoy and Blahna 2001).

The success of any organization lies not in its physical assets, but in the intelligence, skills, and experience of its workforce. Managing the human intellect and harnessing the passion and capabilities of workers into useful products and services is becoming the critical managerial skill of our age. Although there has been significant interest in worker creativity, innovation and learning, there has been little attention paid to managing the professional intellect of knowledge workers. The same can be said about effective supervision of older workers. When fully engaged, older workers offer their organizations four competitive advantages: (1) cognitive knowledge (know-what); (2) advanced professional skills (know-how); (3) systems knowledge (know-why) and (4) self-motivated creativity (care-why) (Anderson et al. 1996). It is for these reasons that human resource strategists have to design and deliver innovative ways to keep older workers productive, satisfied and motivated. However, it is a two-way street: older workers must be willing to improve their technological skills by further education and training.

America needs older workers to compete in the new emerging global economy. It is the responsibility of human resource managers to "retain," "retool," and "reengage" older workers with insightful and ethical policies and procedures, such as those discussed in this chapter.

References

AARP. (2000). *American business and older employees survey*. Washington, DC: AARP.

Amabile, T. (1999). How to kill creativity. In *On breakthrough thinking* (pp. 1–26). Boston: Harvard Business Review.

Anderson, P., Finkelstein, S., & Quinn, J. B. (1996). *Managing professional intellect: Making the most of the best* (pp. 181–205). Boston: Harvard Business Review.

Armstrong-Stassen, M., & Cattaneo, J. (2010). The effect of downsizing on organizational practices targeting older workers. *The Journal of Management Development, 29*, 344–363.

Ashbaugh, D. L., & Fay, C. H. (1987). The threshold for aging in the workplace. *Research on Aging, 9*, 417–427.

Bond, J., Galinsky, E., & Brownfield, E. (2005). *National study of employers*. New York: Families and Work Institute.

Brainard, K. (2002). Phased retirement overview: Summary of research and practices. *National Association of State Retirement Administrators.*

Byham, W. (2008, April). Flexible: Strategies for retaining select employees beyond normal retirement age. *Training & Development (T & D)*, 34–37.

Chandler, Gray & Christmas, Inc. (2010). *Despite high unemployment, older workers gain positions: Older workers join, succeed in job search.* Retrieved from http://www.challengegray.com/press/PressRelease.aspx?PressUid=135

Christensen, K. (2005). *Getting the fit right: Flexible work options* (Brief 3). Retrieved from Boston College Sloan Center of Aging & Work website: http://www.bc.edu/agingandwork

Close, Ellis (2009, November 9). Fired is the New Retired. *Newsweek, 154*, 26.

Crampton, S., Hodge, J. & Mishra, J. (1997, Summer). Transition – Ready or not: The aging of America's workforce. *Public Personnel Management, 25*(2), 221–234.

Dooley, K. (2004, March/April). How to avoid being named in an age discrimination case. *AFP Exchange, 24*, 40–41.

Dychtwald, K., Erickson, T., & Morrison, B. (2004, Summer). It's time to retire retirement. *Public Policy & Aging Report, 14.*

Gale, S. (2011). *Phased Retirement*. Retrieved from http: www.forcework.com/section/benefits-compensation/feature/phased-retirement
Gross v. FBL Financial Services, Inc. (2008). Retrieved from http://www.oyez.org/cases/2008-2009
Harkin, T. (2010). *Ensuring fairness for older workers*. Retrieved from https://secure.acslaw.org/taxonomy/term/1888
Hedge, J. W. (2008). Strategic human resource management and the older worker. *Journal of Workplace Behavioral Health, 21*, 109–123.
Hedge, J. W., Borman, W. C., & Lammlein, S. L. (2005). *The aging workforce: Realities, myths, and implications for organizations*. Washington, DC: APA Books.
Helson, R., & Cate, R. A. (2007). Late middle age: Transition to the third age. In J. James & P. Winks (Eds.), *The crown of life: Dynamics of the early postretirement period* (pp. 83–102). New York: Spring Publishing Company, LLC.
Hutchens, R. (2003). *The Cornell study of employer phased retirement policies: A report on key finds*. Ithaca: School of Industrial and Labor Relations at Cornell University.
James, J., Swanberg, J., & McKechnie, S. (2007). *Responsive workplaces for older workers: Job quality, flexibility and employee engagement* (Brief 11). Retrieved from Boston College Sloan Center of Aging & Work website: http://www.bc.edu/agingandwork
Leadership & Learning (n.d.). Retrieved from htttp://www.ge.com/company/culture/leadership_learning.html
Kaye, B., & Cohen, J. (2008, April). Safeguarding the intellectual capital of baby boomers: When transitioning boomers into retirement, loyalty begets loyalty. *Training & Development (T & D)*, 30–33.
Kahn, W. (1990). Psychological conditions of personal engagement and disengagement at work. *Academy of Management Journal, 33*, 692–724.
Klatt, B. (1999). *The ultimate training workshop handbook*. New York: McGraw-Hill.
Lewis, J. (2009, March). *Best practices for avoiding age discrimination lawsuits and claims in RIFs*. Retrieved from http://www.jacksonlewis/legal/updates
Leonard, B. (2011, January). When HR goes bad. *HR Magazine*, pp. 30–32.
Maurer, T., & Rafuse, N. (2001). Learning, not litigating: Managing employee development and avoiding claims of age discrimination. *The Academy of Management Executive, 15*, 110–121.
McCann, L., & Osborne, T. (2004, Spring). Forced ranking and age-related employment discrimination. *Human Rights: Journal of the Section of Individual Rights & Responsibilities, 31*, 6–10.
McEvoy, G., & Blahna, M. J. (2001). Engagement or disengagement? Older workers and the looming labor shortage. *Business Horizons, 44*, 46–52.
Moen, P., Erickson, W., Argarwal, M., Fields, V., & Todd, L. (2000). *The Cornell retirement and wellbeing study*. Ithaca: Bronfenbrenner Life Course Center at Cornell University.
Norris, F. (2011, January 14). Older workers are keeping a tighter grip on jobs. *New York Times*. Retrieved from http://www.nytimes.com/2011/01/15/business/15charts.html
Office of Technology Assessment (OTA). (1990, September). *Worker training: Competing in the new international economy*. Washington, DC: OTA.
Paine, L.S. (1994, March–April). Managing for organizational integrity. *Harvard Business Review*, 106–117.
Parkinson, D. (2002). *Voices of experience: Mature workers in the workplace* (Research Report). New York: The Conference Board.
Penner, R. G., Perun, P., & Steuerle, E. (2002). *Legal and institutional impediments to partial retirement and part-time work by older workers* (pp. 1–5). Washington, DC: Urban Institute.
Pew Research Center. (2010). *Millennials: A portrait of generation next: Confident. ConnectedOpen to change*. Washington, DC: Pew Research Center.
Pitt-Catsouphes, M. & Smyer, M. (2005). *Older workers: What keeps them working* (Brief 1). Retrieved from Boston College Sloan Center of Aging & Work website: http://www.bc.edu/agingandwork
Putnam, H. (2009, November 3). *Turbulence is inevitable...misery is optimal: Ethics and integrity are your greatest assets in good times*. Waltham: The Raytheon Lectureship in Business Ethics at Bentley University.

Ranstad Corporation (2008, October 15). *World of work 2008*. Retrieved from http://www.us.ranstad.com

Reid v. Google, Inc. (2010). *California courts*. Retrieved from http://courtinfo.ca.gov/opinions/revpub/H029602.DOC

Reid v. Google: California Supreme Court Creates New Obstacles to Summary Judgment for Employers. (2010). Retrieved from http://californiaemploymentlaw.foxrothschild.com/tags/reid-v-google

Reidenbach, R. E., & Robin, D. P. (1991). A conceptual model of corporate moral development. *Journal of Business Ethics, 10*, 273–284.

Robinson, A. G., & Stern, S. (1998). *Corporate creativity*. San Francisco: Berett-Koehler.

Rosen, B., & Jerdee, T. (1976). The nature of job-related age stereotypes. *Journal of Applied Psychology, 61*, 180–183.

Roth, D. (2011, January 27). Davos Journal 2011: Why companies need to start hiring older workers. *CNNMoney.com*. Retrieved from http://features.blogs.cnn.com/2011/01/27

Saks, A. M. (2006). Antecedents and consequences of employee engagement. *Journal of Managerial Psychology, 21*, 600–619.

Samuelson, R. J. (2010, December 20). The flight from risk. *Newsweek* 20.

Seneca, J., & Hughes, J. W. (2009). *America's demographic tapestry*. New Brunswick: Rutgers University Press.

Sullivan, J. D. (2009). The moral compass of companies: Business ethics and corporate governance as anti-corruption tools. *Global Corporate Governance Forum, 7*, 1–55.

Technology Trends: The changing workplace. (2003, Fall). *SeniorNet*. Retrieved from http://seniornet.org/php?

The Aging Workforce in America (n.d.). Retrieved from http://www.imec.org/imec.nsf?All?Aging_Workforce?

White, T. (n.d.). *Ethics incorporated: How America's corporations are institutionalizing moral values*. Retrieved from Loyola Marymount University Center for Ethics and Business website: http://imu.edy/Assets/Colleges

Workplace Flexibility. (2010). Retrieved from Boston College Sloan Center on Aging & Work website: http://www.bc.edu/research/aging/about/workFlexibility.html

Yeatts, D. E., Folts, W. E., & Knapp, J. (2000). Older workers' adaptation to a changing workplace: Employment issues for the 21st century. *Educational Gerontology, 26*, 566–582.

Zambito, T. (2011, January 21). Ellen Strickman, 80 says she was fired by Home Depot because of her age. *New York Daily News*.

Zeytinoglu, I., Cooke, G. B., & Harry, K. (2007, March). *Older workers and on-the-job training in Canada: Evidence from the WES data* (SEDAP Research Paper No. 179). Retrieved from McMaster University website: http://socserv.macmaster.ca/sedap

Chapter 8
Counseling Older Workers Confronting Ageist Stereotypes and Discrimination

Margo A. Jackson

As people live longer and work longer, the prevalence of older adults in the workforce has increased. Ever greater numbers of these workers face an economic recession that has exacerbated financial stress and sharply raised the unemployment rate (Butler 2008; Sok 2010; U.S. Bureau of Labor Statistics 2008).

Research suggests that older workers contribute many positive attributes such as knowledge, productivity, work ethic, stability, honesty, dependability, and mentoring (Berger 2009; Hall and Associates 1996). Despite significant life and work experiences, pervasive ageism operates in overt and covert ways to discriminate against older workers in hiring and workplace practices (Bendick et al. 1999; Genevay 2000; Noonan 2005). This chapter is written from the author's perspective as a vocational counseling psychologist in the U.S. It aims to provide helping professionals with theoretical and practical information for counseling older workers who confront ageist discrimination.

Work discrimination has been defined as "unfair and negative treatment of workers or job applicants based on personal attributes that are irrelevant to job performance" (Chung 2001, p. 34). Ageism has been defined as "systematic stereotyping of and discrimination against older people because they are old, just as racism and sexism accomplished this with skin color and gender" (Butler 1989, p. 139). Some laws and employment policies prohibit age discrimination in hiring and workplace practices; for example, the Age Discrimination in Employment Act of 1967 (U.S. Equal Employment Opportunity Commission 2011) provides some protections for people who are age 40 or older. Despite these laws and policies, ageism in hiring and

M.A. Jackson (✉)
Division of Psychology and Educational Psychology, Graduate School of Education, Fordham University School of Education,
113 West 60 Street, 8th floor, New York, NY 10023, USA
e-mail: mjackson@fordham.edu

P. Brownell and J.J. Kelly (eds.), *Ageism and Mistreatment of Older Workers: Current Reality, Future Solutions*, DOI 10.1007/978-94-007-5521-5_8,

workplace practices is often subtle and insidious, particularly for people approaching or beyond a traditional retirement age of 65 (McVittie et al. 2003). Genevay (2000) reported the following example:

> Joan was a sixty-one-year-old bookkeeper who had worked in an agency for fifteen years. She was well liked and was an excellent bookkeeper. A new executive, hired to cut costs, decided to terminate Joan because her salary and benefits were more costly to the agency than a new and less experienced employee's would be.
>
> Joan had arthritis and occasionally complained about her fingers. The executive never used written or verbal language that could contribute to an age discrimination suit, but he frequently commiserated with Joan about how hard it must be to "keep up with her work" with arthritic fingers.
>
> He kept his voice empathetic and his words benign, but Joan was fearful and he was able to force her out through subtle intimidation. Joan lost Social Security and other retirement benefits. She eventually found another job with lower salary and fewer benefits, but she had to work many more years in order to be able to afford retirement. (pp. 258–259)

Effects of Negative Ageist Stereotypes on Older Workers

Workplace discrimination against older workers is often based on systemic negative stereotyping. The work discrimination in the case of Joan (Genevay 2000) involved an employment supervisor using an ageist stereotype (namely, that an aging worker cannot "keep up with her work") and an abuse of power to subtly demean and then dismiss the older worker in order to replace her with a less expensive employee in order to gain a short-term profit. Yet even in cases of ageist work discrimination based on less conscious or unintentional stereotyping, the negative economic and psychosocial effects on older workers are substantial. Negative stereotypes commonly held by employers about older workers include beliefs that older workers are inferior to younger workers and are "considered to be less adaptable, less interested in technological change, less trainable, less physically strong, less likely to be promoted, less ambitious, less energetic, less healthy, less creative, less mentally alert, and less flexible, and to have a less functional memory" (Gringart et al. 2008, p. 752). Although empirical evidence has disputed most of these ageist stereotypes, employers continue to believe them. The negative age bias of employers has been linked to reduced opportunities for hiring, training, promotion, and retention of older workers (Chiu et al. 2001).

This author proposes that negative age-based stereotypes may impair the performance and health of older workers. *Stereotype threat* refers to the expectation that individuals will be judged on the basis of a negative stereotype about their social group membership rather than their actual performance and potential (Steele et al. 2002). A substantial body of empirical evidence with other groups vulnerable to negative stereotyping (e.g., groups defined by gender, race, ethnicity, and socioeconomic status) has established that individuals' performance can be harmed by the awareness that they may be judged on the basis of negative stereotypes, regardless of the accuracy of the stereotypes and whether the individuals themselves believe the stereotypes are true (Block et al. 2011). Block et al. suggested that long-term exposure to workplace stereotype threat and discrimination not only has negative effects on work performance but also on workers' physical and mental health.

Meaning of Work in Later Life Transitions

Counselors can be helpful to older workers by exploring strategies for coping with stereotype threat and managing ageist work discrimination. As a first step, however, counselors should assess the meaning of work for the individual in the developmental context of his or her life transitions. Great diversity characterizes the needs, desires, and goals of people in old age, more so than in any other developmental stage of life (Erikson 1963). Noonan (2005) explored the meaning of work for older adults by interviewing a sample of individuals aged 55–77 who were employed or actively seeking employment. The diversity in work-related concerns and experiences of these older adults included confirming, renegotiating, and reforming their work role identities at this stage in their lives. The issues reported by these older workers included: experiencing age discrimination in hiring and premature termination, with demoralizing relational disruptions; financial concerns; wanting part-time work for reasons of health, balance, or generative goals; seeking structure and meaningful activity through work; and identifying specific work environments and conditions consistent with their evolving identities and life transitions. Noonan (2005) interpreted these findings using stages of career development proposed by Super (1980): establishing, maintaining, and disengaging. These stages reflect the diversity of experiences of older workers in later-life work transitions. In the context of their particular life stories and circumstances, the older workers whom Noonan interviewed reported disengaging from their work, maintaining their work roles in the face of age discrimination and financial concerns, and beginning to establish entirely new work lives.

Blustein (2006) proposed that there are three fundamental human needs that working has the potential to fulfill in people's lives: survival (money and access to resources for food, shelter, and clothing), relatedness (access to social support and relational connections), and self-determination. Blustein referred to self-determination as the motivation to engage in working, because of, or regardless of, job satisfaction, when one values the outcomes of his or her work efforts and has access (on or off the job) to supportive contextual conditions that foster autonomy, competence, and relatedness. Older workers share these fundamental human needs.

Ageist work discrimination violates article 23 of the Universal Declaration of Human Rights, adopted by the United Nations General Assembly in 1948: "Everyone has the right to work, to free choice of employment, to just and favorable conditions of work and to protection against unemployment" as noted by Powell (2010, p. 654). Thus, to ensure the human rights of older workers, advocates should address ageist stereotype threat and discrimination.

Counseling Older Workers Confronting Ageist Stereotypes and Discrimination: Coping Strategies

Helping professionals such as counselors, social workers, psychologists, human service and health care providers, and Employee Assistance Program professionals have an important role to play in helping older workers. They can provide, through counseling interviews, first and foremost, invaluable and foundational empathic understanding by nonjudgmentally eliciting, reflectively listening, and accurately affirming clients' perceived experience of ageist work discrimination. A therapeutic relationship is further built on assessment of the specific meanings of work for each client. This assessment must include the client's current life and developmental transitions, as well as the client's assets and sources of support. In addition, the counselor should identify the client's goals in confronting challenges and barriers. These goals will then lead the counselor and worker to develop constructive coping strategies. In the following discussion, this author integrates two conceptual models and empirical findings that counselors might use to assess and explore with older adults a range of helpful strategies for managing ageist work discrimination.

In a study of the strategies used by older adults for managing age discrimination in their search for employment, Berger (2009) found they used two types of tactics to offset negative stereotypes: *counteractions* and *concealments*. The majority of the older adult participants tried to counteract prospective employers' ageist stereotypes by maintaining their skills and keeping up-to-date in training. When confronted with age discrimination in their job searches, most of the participants reported they changed their work-related expectations regarding the goals, type, or geographic location of employment they were seeking. Concealment strategies used to deal with age discrimination included altering their resumes, modifying their physical presentation to appear more youthful, using current buzzwords in their field, and reframing responses to age-discriminatory questions. Berger noted that despite the structural barriers of age discrimination, participants in her study appeared to maintain agency or control in the job search process by actively negotiating strategies to offset ageist stereotypes. Furthermore, she asserted that ageist work discrimination must be fought on a broader societal level, for example, by policy makers and employers working together to include age in their diversity strategies for recruitment and retention. Taking individual clients' needs into account, counselors and older workers should consider counteraction strategies for managing ageist work discrimination. Counselors should also discuss with clients the costs and benefits of concealment strategies.

Although counteraction and concealment strategies for managing ageist work discrimination may help older adults maintain agency, or remain centered when faced with potentially negative stereotypic responses, in the job search process (Berger 2009), older workers need a range of coping strategies to maintain their performance and health while negotiating long-term exposure to workplace stereotype threat and discrimination. Drawing from social, organizational, and counseling psychology literatures, Block et al. (2011) proposed a conceptual model of coping strategy responses to long-term stereotype threat in the workplace. Although the Block et al. model focused more on work discrimination experiences of women and people of color, this author suggests counseling applications might be used more widely in remediation, prevention, and advocacy with older adults confronting work discrimination. The model posits three response sets to stereotype threat (which may occur in stages): fending off the stereotype; discouraged by the stereotype; and resilient to the stereotype. Within each response set, Block et al. noted several coping strategies that an individual may employ, with varying costs and benefits.

Fending Off the Stereotype

Individuals in this phase work vigorously to demonstrate that the stereotype does not apply to them and try to show that they are not a typical member of their social identity group (Block et al. 2011). *Invigoration*, or overcompensating by working harder to an extreme, is one strategy that individuals may use to fend off the stereotype. For example, an older adult might use this strategy to counter the stereotype that older workers are less energetic than younger workers. Although one benefit of this strategy may be high levels of work productivity, there may be corresponding negative consequences for workers' mental and physical health.

In work environments, people inevitably fall short of their goals at times. *Internal attributions* for failure to achieve one's goals is a strategy often used by members of stigmatized groups to fend off stereotypes at work (Block et al. 2011). Particularly when discrimination is ambiguous, stereotyped individuals tend to attribute negative outcomes to their own inadequacies rather than to discrimination (Ruggiero and Taylor 1997). For example, in the case of Joan cited earlier (Genevay 2000) regarding her failure to meet the new executive's expectations of her work performance, she may have attributed this to her inability to "keep up" due to occasional arthritis, rather than to his subtle yet frequent ageist intimidation intended to force her out of the job. Block et al. posited that internal attributions for fending off the stereotype may function as a protective measure for individuals to retain some sense of control over their own performance rather than to feel helpless in the face of discrimination. Over time, however, individuals faced with repeated discrimination who rely on internal attributions for failure to meet goals may become discouraged and depressed.

Two other coping strategies for fending off stereotypes at work are *identity bifurcation*, whereby individuals reject aspects of their identity that are subject to discrimination, and *assimilation*, whereby individuals attempt to assume the characteristics of the more positively regarded social group (Block et al. 2011). For example, the concealment strategies used to deal with age discrimination by the older adults in Berger's (2009) study included altering the presentation of their work history and physical appearance to de-emphasize their age and seem more youthful.

In counseling older adults confronting work discrimination, potential costs and benefits of these strategies should be considered. There may be potential advantages in effectively communicating and marketing one's skills developed from a substantial portfolio of work experience. Likewise, older workers may effectively debunk ageist stereotypes by highlighting their energy, strength, and adaptability in their physical presentation and demeanor in the workplace. On the other hand, there may be potential costs to using identity bifurcation and assimilation strategies (Block et al.), or concealment strategies whereby older workers fend off the stereotype by hiding aspects of themselves to demonstrate how they are not a "typical" member of their age group. Rejecting their social identity age cohort group may be a stressful and unsustainable strategy over the long term.

Discouraged by the Stereotype

Block et al. (2011) posited that over an extended period of time, when fending strategies are met with persistent discriminatory encounters whereby they are evaluated based on the stereotype, individuals often respond by becoming discouraged. One coping strategy used in this response set is psychological *disengagement* of one's self-esteem from one's work performance (e.g., by devaluing or discounting performance feedback from one's supervisor). Other coping strategies include making *external attributions* to discrimination rather than to one's ability when failing to achieve goals; *anger* directed toward the perpetrator or oneself; and psychological or behavioral *withdrawal* from the workplace, resulting in negative work attitudes, tardiness, absenteeism, or turnover.

Albeit understandable responses, potential consequences of older workers using these coping strategies may include poorer health outcomes over time (Block et al. 2011), disrupted interpersonal relationships (on and off the job), and declines in work performance that are then used by employers to justify demotion or termination. Furthermore, risk for depression may be heightened for older adults who commonly deal with cumulative losses from stressors in many other life domains (e.g., deaths of friends and loved ones, chronic physical health conditions, caregiving for dependent adult family members) (Rainsford 2002). Fortunately, counseling may be particularly helpful for older workers experiencing this phase of coping responses to workplace stereotype threat. For example, research evidence documents that older adults respond well to a variety of forms of counseling, psychotherapy, and psycho-educational interventions for depression and anxiety, although less is known about the efficacy of counseling interventions with ethnic minority older adults (American Psychological Association 2003). In addition, counselors can empathically affirm the anger that older workers experience in response to persistent ageist discrimination and help them to explore more effective strategies to redirect or channel their anger toward constructive goals.

Resilient to the Stereotype

According to Block et al. (2011), many people who are subject to workplace stereotype threat, yet remain engaged in their work and satisfied with their careers, use coping strategy responses characterized by a recovery stage. Individuals who respond to stereotype threat with resilient strategies demonstrate a capacity to acknowledge, yet recover, from work discrimination setbacks and redirect their energy in ways that foster further development. *Challenging negative group stereotypes* by directly confronting and educating members of the dominant group about the effects of their misconceptions is one resilient coping strategy used in this response set. For example, an older worker might assertively and respectfully discuss with her supervisor how the decision to exclude her from consideration for a training program in new technological tools seemed to be based on the misconception that "old dogs can't learn new tricks." Rather, this supervisor should base training decisions on concrete evidence of the worker's sincere interest and frequent participation in opportunities to develop her technological competencies.

Another resilient coping strategy to workplace stereotype threat is *positive distinctiveness,* whereby individuals communicate the favorable attributes of their social identity group (Block et al. 2011). For example, in their workspaces, older workers might display their awards for mentoring accomplishments, lifetime achievement, or other professional distinctions. *Collective action* is a resilient coping strategy of joining with others to promote change in the workplace by raising awareness of the influences of negative stereotype threats and shifting to more inclusive institutional and sociocultural norms and access to resources (Block et al. 2011). Forty Plus Clubs (non-profit career development organizations) and AARP (American Association of Retired Persons) are examples of employee networks and affinity groups that may provide social support, developmental opportunities, and advocacy for older workers.

Finally, Block et al. proposed *redefining criteria for success* at work as a resilient coping strategy, for example in bypassing barriers of workplace stereotype threat and starting one's own business. Entrepreneurial opportunities for designing, manufacturing, and marketing products to the growing number of older adults may favor older workers (Hall 1996). Yet, this strategy requires access to resources to start a business. These resources include capital, and advisors such as those available at local branches of the U.S. Small Business Administration.

Benefits to using the resilient coping strategies of challenging stereotypes and conveying positive distinctiveness include enhancing individuals' self-esteem by demonstrating their competence in the face of discrimination (Block et al. 2011). A potential risk for the individual who repeatedly challenges the negative stereotype of the group is that he or she may be labeled as a troublemaker and isolated by the dominant group. Thus, this response may also preclude the stereotyped group from achieving more equitable treatment. Group-level strategies of engaging in collective action, although disruptive and not without risks, can effectively result in more equitable systemic changes that benefit both the disadvantaged and dominant groups in the long term (Block et al. 2011). Counselors can help older workers explore and evaluate a range of resilient coping strategies to address their particular experiences and concerns in responding to long-term workplace stereotype threat and age discrimination.

Yet another model—Y. B. Chung's (2001) Work Discrimination and Coping Strategies Model—can be used by counselors in remediation, prevention, and advocacy with older adults confronting work discrimination. Although Chung's model was developed specifically for lesbian, gay, and bisexual (LGB) workers, he and his colleagues suggested that future researchers investigate how the model might be applied with other forms of work discrimination. For example, the model might be applied to age discrimination, and could be used to explore older workers' coping strategies (Chung et al. 2009).

Chung et al. (2009) validated multiple coping strategies used by LGB workers to manage discrimination incidents they anticipated or encountered. These included: (a) *using internal coping strategies* such as silence, avoidance, self-talk, and overcompensating in their work performance; (b) *seeking sources of social support* from a partner, friends, family members, coworkers, and mental health providers; and (c) *taking action to confront the discriminatory act*, such as talking directly with the offender, requesting supervisor intervention, submitting a grievance to human resources, pursuing legal counsel or litigation, defying the discriminatory policy, and seeking media attention to publicize the discriminatory act and influence the employer. These coping strategies can be applied in counseling older workers. In particular, the coping strategies of seeking sources of social support and taking action to confront the discriminatory act may be useful to explore with older workers confronting ageist stereotypes and discrimination.

Furthermore, counselors might help older workers advocate for reasonable workplace accommodations targeted to their physical and psychosocial needs. For example, there are increasing physical limitations as people age, such as reduced hearing and vision (Rainsford 2002). Just as reasonable workplace accommodations are required for employees with other physical impairments (for example, those required by the Americans with Disabilities Act of 1990, as amended), there should likewise be support for older workers to develop helpful adaptations in the workplace to facilitate their inclusion and work productivity.

Conclusion

Employment and workplace discrimination based on negative ageist stereotyping against older adults is prevalent (Chiu et al. 2001; Gringart et al. 2008). Challenges may be compounded for older workers who are members of other social identity groups vulnerable to stereotype threat and work discrimination (e.g., groups defined by gender, race/ethnicity, socioeconomic status, or sexual orientation). In addition, counselors are not immune to ageist biases (Rainsford 2002). Thus, counselors have an ethical responsibility to consciously examine their own assumptions and continually monitor the influence of ageist (and other negative) stereotypes in their counseling with older adults confronting work discrimination.

Although empirical evidence has disputed commonly held ageist stereotypes against older workers, negative age bias has been shown to reduce opportunities for hiring, training, promotion, and retention of older workers (Chiu et al. 2001). Workplace stereotype threat (the expectation that older workers will be judged by ageist stereotypes instead of their performance and potential) may effectively impair not only their work productivity, but also the physical and mental health of older workers. This chapter included a discussion of issues for counseling older workers confronting ageist stereotypes and discrimination in employment and the workplace. The author presented a range of potentially helpful counter-strategies that counselors can use to help older workers.

References

American Psychological Association. (2003). *Guidelines for psychological practice with older adults*. Washington, DC: American Psychological Association.

Americans with Disabilities Act of 1990, as amended (n.d.). Retrieved May 3, 2011 from http://www.ada.gov/pubs/adastatute08.htm

Bendick, M., Jr., Brown, L. E., & Wall, K. (1999). No foot in the door: An experimental study of employment discrimination against older workers. *Journal of Aging & Social Policy, 10*, 5–23.

Berger, E. D. (2009). Managing age discrimination: An examination of the techniques used when seeking employment. *Gerontologist, 49*, 317–332.

Block, C. J., Koch, S. M., Liberman, B. E., Merriweather, T. J., & Roberson, L. (2011). Contending with stereotype threat at work: A model of long-term responses. *The Counseling Psychologist, 39*, 570–600. doi:10.1177/0011000010382459.

Blustein, D. L. (2006). *The psychology of working: A new perspective for career development, counseling and public policy*. Mahwah: Erlbaum.

Butler, R. N. (1989). Dispelling ageism: The cross-cutting intervention. *The Annals of the American Academy of Political and Social Science, 503*, 138–147.

Butler, R. N. (2008). *The longevity revolution: The benefits and challenges of living a long life*. New York: Public Affairs.

Chiu, W. C. K., Chan, A. W., Snape, E., & Redman, T. (2001). Age stereotypes and discriminatory attitudes towards older workers: An East-West comparison. *Human Relations, 54*, 629–661.

Chung, Y. B. (2001). Work discrimination and coping strategies: Conceptual frameworks for counseling lesbian, gay, and bisexual clients. *Career Development Quarterly, 50*, 33–44.

Chung, Y. B., Williams, W., & Dispenza, F. (2009). Validating work discrimination and coping strategy models for sexual minorities. *Career Development Quarterly, 58*, 162–170.

Erikson, E. (1963). *Childhood and society* (2nd ed.). New York: W. W. Norton.

Genevay, B. (2000). There is life after work: Re-creating oneself in the later years. In N. Peterson & R. C. González (Eds.), *Career counseling models for diverse populations: Hands-on applications by practitioners* (pp. 258–269). Belmont: Wadsworth/ThomsonLearning.

Gringart, E., Helmes, E., & Speelman, C. (2008). Harnessing cognitive dissonance to promote positive attitudes toward older workers in Australia. *Journal of Applied Social Psychology, 38*(3), 751–778.

Hall, D., & Associates. (1996). *The career is dead – Long live the career*. San Francisco: Jossey-Bass.

McVittie, C., McKinlay, A., & Widdicombe, S. (2003). Committed to (un)equal opportunities?: 'New ageism' and the older worker. *British Journal of Social Psychology, 42*, 595–612.

Noonan, A. E. (2005). "At this point now": Older workers' reflections on their current employment experiences. *International Journal of Aging & Human Development, 61*, 211–241.
Powell, M. (2010). Ageism and abuse in the workplace: A new frontier. *Journal of Gerontological Social Work, 53*, 654–658.
Rainsford, C. (2002). Counseling older adults. *Clinical Gerontology, 12*, 159–164.
Ruggiero, K. M., & Taylor, D. M. (1997). Why minority group members perceive or do not perceive the discrimination that confronts them: The role of self-esteem and perceived control. *Journal of Personality and Social Psychology, 72*, 373–389.
Sok, E. (2010, March). Record unemployment among older workers does not keep them out of the job market. *U.S. Bureau of Labor Statistics*. Retrieved from http://www.bls.gov/opub/ils/summary_10_04/older_workers.htm
Steele, C. M., Spencer, S. J., & Aronson, J. (2002). Contending with group image: The psychology of stereotype and social identity threat. In M. P. Zana (Ed.), *Advances in experimental social psychology* (Vol. 34, pp. 379–440). San Diego: Academic.
Super, D. E. (1980). A life-span, life-space approach to career development. *Journal of Vocational Development, 16*, 282–298.
U.S. Bureau of Labor Statistics (2008, July). *Older workers: Are there more older people in the workplace?* Retrieved from http://www.bls.gov/spotlight/2008/older_workers/
U.S. Equal Employment Opportunity Commission (2011, April 1). *The age discrimination in employment act of 1967*. Retrieved from http://www.eeoc.gov/laws/statutes/adea.cfm

Chapter 9
Legal and Legislative Issues in Protecting Older Adults in the Workplace: An International Perspective

Charmaine Spencer

Betty lost her job at the age of 68, and lost her human rights age and gender discrimination case two and a half years later (Braun 2009).

People heard she was dismissed because she refused to learn how to use a computer; the reality is more complex. The new office supervisor felt Betty was undermining his authority. For months he repeatedly used profanity and called Betty highly derogatory names behind her back and in the presence of others (including a B**, an old F*B*, and a F*B*).

Betty was not the only source of the supervisors' wrath. Once while at work the supervisor wrote an article in which he ridiculed three older front office staff members as "rambling on about nothing". He referred to the office as "a great place for elderly people to come to when they are no longer any use to society." He also wrote in the article "At our company we have a saying 'the equipment is old, and so are the people', so grab your canes and wheelchairs and come on down and have a gay old time at …[our company's] Senior Citizen's Day Care." This was not his only use of harassment. When asked at the hearing why he took away the parking space that Betty had for years, he said it was "because he could".

When Betty brought the actions and ridicule to the attention of the owner, he did nothing about it. One of the other older office staff members retired as soon as he could. Tensions between the new supervisor and Betty increased. Betty's health and work suffered. Betty and the other employees tried to have the supervisor fired without success. Instead Betty lost her job.[1]

[1] Adapted from the facts of *Braun v. Profile Seismic Ltd. 2009 AHRC 3.* Human Rights Panels of Alberta (June 16, 2009) Diane Colley-Urquhart adjudicator.

C. Spencer LL.B., J.D, LL.M. (✉)
Gerontology Research Centre, Simon Fraser University, 2800-515W. Hastings Street, BC V6B 5K3, Vancouver, Canada
e-mail: cspencer@shaw.ca

P. Brownell and J.J. Kelly (eds.), *Ageism and Mistreatment of Older Workers: Current Reality, Future Solutions*, DOI 10.1007/978-94-007-5521-5_9,

At the beginning of the twentieth century, most people in North America worked in a factory or private business, at home, or in the field until their health deteriorated. By mid – century, a fundamental shift had begun: due to increased life expectancy, some people could now retire. Retirement became a by-product of modern society, part of the fabric of labour market organization, and age 65 became the generally accepted retirement age in North America.

As advantageous as this was for many workers, it also led to a "conveyor belt" expectation: workers would enter the labour force in their teens or twenties, move along at a uniform rate, and then, like assembly line parts, drop off the labour conveyor belt at a defined "natural" point, to make room for younger workers behind them (Spencer 2003).

The public, employers and policy makers often assume there is only a fixed amount of work available within an economy. Economists sometimes refer to these as "boxed economy" assumptions and the "lump of labour fallacy" (e.g., Cahill 2011). However, in this zeitgeist, older workers have been expected to leave the workforce at a defined age, and this has been seen as socially and morally justifiable. In *McKinney v. University of Guelph* 1990 the Supreme Court of Canada ruled that retirement was a by-product of modern society and that 65 had now become generally accepted as the "normal age of retirement". In line with this perspective was the idea that old workers had a natural responsibility to "make way" for younger workers (at 294–95). During economic downturns, they were expected to "sacrifice" themselves for the benefit of unemployed young workers, in order to redistribute the scarce resource of work (Lowenstein and Doron 2008). It was a view of the world that valued the young over the old, pitted generations against each other, and subtly, and not so subtly, delivered messages that pushed older workers out of the workforce permanently or treated them as easily expendable.

This chapter considers the connections and distinctions among the terms *ageism, age discrimination, workplace violence, workplace bullying, harassment, and mistreatment.* It examines the limited available evidence of age as a factor in workplace bullying. It looks at the development, capacity and efficacy of laws in Canada, United States and Europe to protect older workers from discrimination and harassment in the workplace. It considers not only the laws, but also the social and legal environment in which they are applied. In this context, it is important to consider not only age *per se* as affecting older workers and leading to discrimination, harassment or bullying in the workplace. There also may be a wider constellation of issues associated with being an older worker that can lead to older workers' harassment in the workplace. These issues include job sector, lifelong financial insecurity, being widowed, supporting a disabled spouse or caring for grandchildren (Institute for Women's Policy Research 2012).

Ageism, Age Discrimination, Workplace Violence, Workplace Bullying, Harassment, and Mistreatment.

When Robert Butler (1975) first coined the term ageism in the late 1960s, he framed it in terms of positive or negative attitudes and beliefs about older compared to younger people. Today ageism refers to negative attitudes and beliefs about older people and negative behaviours towards them. Ageism includes both individually held personal beliefs, and societal beliefs about older people. Although some ageism may manifest as open hostility, it commonly comes in more subtle forms that effectively marginalize older persons, and that treat them as less valued than younger adults.

Ageism can operate in any sphere, including the workplace. It can occur in any phase of employment and in any aspect of employment, including before hiring, working conditions, pay and benefits, and termination. Although some ageism surfaces as dislike or intolerance of older people, the more common forms of ageism and discrimination based on age rest on unsupported general assumptions about the effect of age on ability.

The Ontario Human Rights Commission (2007) noted that older adults can feel the effect of age discrimination when trying to re-enter or remain in the workforce. Older prospective workers are likely to encounter not-so-subtle comments such as, "Are you sure you can handle this job? It takes a lot of energy and enthusiasm, and we are looking for someone with career potential." While in the job, older workers may be told, "You don't need this training program. At your age, what would the benefit be?" An employer may assume that an older worker is under-productive, prone to frequent absences, too difficult to train and "riding it out to retirement" (Ontario Human Rights Commission 2008). Ageism can occur in hiring practices, when human resource personnel consciously or subconsciously apply age limits to older applicants. Age discrimination can occur unintentionally when "neutral" performance tests used with older applicants or workers do not account for individual variations in visual acuity, hearing, testing speed or learning style.

As a result of ageism, older workers can and do experience workplace bullying and harassment. These terms refer primarily to psychological harm inflicted against a worker in the workplace by fellow workers, managers or others. The intent is to demean the person, although this can also be backed by threats and physical violence. According to the Canada Safety Council (n.d.):

> A workplace bully subjects the target to unjustified criticism and trivial fault-finding. In addition, he or she humiliates the target, especially in front of others, and ignores, overrules, isolates and excludes the target person. Regardless of specific tactics, the intimidation is driven by the bully's need to control others.

Workplace bullying may involve an individual using power and control over others. In some cases, the targets are of a specific age, race or other characteristic. About one third of all bullying incidents involve what is considered an identifiable characteristic. In many cases, the bully simply goes after whoever is seen as somehow threatening their position, or simply vulnerable. The target may be socially weaker or have lower social status, or simply be non-confrontational. The individual bully may work alone or bring others to participate in the harassment. Workplace bullying creates a hostile work environment, not only for the targeted individual, but also for others.

The objective of workplace bullies is to intimidate or terrorize the person or to have the person quit. The effects on the employees are significant. Signs and symptoms may include: grievances by employees against their manager; declining work performance of dedicated and hard-working employees; increased stress and tensions between staff in a unit; poor morale; reported fear of a co-worker by other workers; individual symptoms of depression and increased absenteeism in a department/unit

(Ontario Safety Association for Community and Health Care 2009). For the target, the workplace bullying or harassment becomes a job stressor, negatively affecting the person's health and leading to frequent absences from work. This then becomes a reason to lay off the targeted worker. This is particularly problematic for older workers because they often are blamed for high health care costs.

There are several common characteristics of workplace bullying: it tends to involve a pattern of negative behaviour or a pattern of negative communication and there is often a power disparity between the target and the bully. Workplace bullying tends to go on for an extended time and it typically escalates (Lutgen-Sandvik et al. 2009). It causes significant harm and the target typically believes the bully is acting intentionally.

The available statistics on bullying identify its gendered aspect (84% of bullied employees are female); longevity (on average, targets endure bullying for almost 2 years before filing a complaint); and low level of sanctions (only a small percentage of workplace bullies, fewer than 13%, are ever punished or terminated) (Namie 2003).

Workplace violence is a more specific term that refers to any variety of physical harm (including assaults) against a worker in the workplace. The harm may come from fellow workers, from managers, from others (spouse, clients, customers), or even strangers.

Unemployment, Discrimination and Workplace Vulnerability

The relationship between age and workplace displacement is complex. According to the U.S. Department of Labor, the jobless rate for people 55 and older rose to 7.3% in August, 2010, from 3% before the recession began, making 2008–2010 the longest spell of high unemployment that workers in this age group had experienced in 60 years. Of this age group, African Americans and Hispanics were disproportionately affected. Hispanic men were the worst off, with an unemployment rate of almost 13% for individuals aged 55 and older (Schwartz 2010).

In 2009, 80.4% of Canadians aged 25–44 had jobs. This age group was more likely to be working than other age groups. The employment rate was 55.3% for youth aged 15–24 and 70.3% for individuals aged 45–64. Among people aged 65 years or older, 10.1% had jobs in 2009 (Human Resources & Skills Development Canada n.d.).

The labour market and employment percentages often cited for the 55 and older age group may be an erroneous representation of reality. Some people in their late fifties and early sixties may describe themselves as "retired", rather than use the stigma-laden term "unemployed" when they have been out of work for an extended period of time. Many of these older people may simply "age out" of the labour force and "involuntarily retire" (O'Connell 2005; Szinovacz and Davey 2005).

In Canada and the United States, older workers who lose their jobs spend far more time searching for work and are jobless far longer than workers younger than 55 (Pittman 2010). For example, in Canada, in 2009, workers aged 55–64 remained unemployed longer (28.3 weeks on average) than other age groups (Human Resources & Skills Development Canada, n.d.). The average was 23.4 weeks for people aged 65+, and 18.4 weeks for those aged 25–54. It is not clear if the figures reflect workplace ageism, or differences between older and younger workers' skill levels and transferability of skills between industries.

Worker gender is another factor in worker unemployment, discrimination and vulnerability. For example, during recessions, a greater percentage of men than women lose their jobs, at all ages (Schneider 2011). Women's jobs are less well paid, and the power disparity within the job means that they may be easier prey to different forms of harassment and other victimization from coworkers or employers (Women's Institute for Policy Research 2012; Drolet 2011; Catalyst 2011).

There are likely two different types of harassment in this area: one directed at particular workers and another that is systemic. In the latter type, efforts to push out older people are motivated by labour management or economic issues. For example, in the United States, older workers may be pushed out because employers consider them more expensive to insure and to maintain on the payroll. As the Workplace Bullying Institute (2010a) suggested, "Employers want to drive out the more experienced, typically higher paid, workers" (pp. 1).

In the United States, the available remedies for age discrimination are tested at a greater rate during economic downturns, such as the current one. Age discrimination complaints filed with the Equal Employment Opportunity Commission (EEOC) rose to over 23,200 in fiscal year 2010, up 40% from 2006. Now, age-based claims make up nearly a quarter of all discrimination complaints received by the EEOC (Equal Opportunity Employment Commission 2010). Some Canadian tribunals have reported more age discrimination cases being adjudicated. For example, the Canadian Human Rights Commission (CHRC), which is responsible for the federal public service and federally regulated industries, noted, "Data collected from the Commission's Dispute Resolution Branch suggests that age discrimination has already started to have a significant impact on Canadian workplaces. This past year [2010], 227 of the 853 complaints (27%) accepted cited age as one on the grounds of discrimination" (Canadian Human Rights Commission 2010). Over the past 5 years, the CHRC referred 273 complaints of age discrimination to the Canadian Human Rights Tribunal for formal adjudication. However, almost all were related to mandatory retirement, rather than harassment or other work conditions.

Age as a Factor in Workplace Bullying

Are older workers more likely than other workers to experience workplace bullying? Is age a key dimension of bullying in later life? Is bullying more likely in a recession? If so, how susceptible are older workers to workplace bullying? There is scant evidence on the vulnerability of older workers or extent of the problem among older workers. For example, research to study whether bullying is more likely in a recession has led to mixed results (Workplace Bullying Institute 2010b).

The Workplace Bullying Institute (WBI) found an overrepresentation of older American workers (aged 50 and up) seeking help from their counselling training.

However, at least one study, relying on on-line self-reports from workers, suggested that 30–49 year olds were more likely to report bullying than those who were younger or older (Workplace Bullying Institute 2010b). There were, however, limitations to this study. The on-line survey did not include workers older than 64. In addition, 30–49 year olds are much more likely than older workers to go on-line. (This is the case in Canada, where about one half of people aged 65–74 use the Internet compared to 80% of those aged 45–64 (Statistics Canada 2011).

Protections through Anti-age Discrimination Laws in North America

Anti-age discrimination legislative remedies passed in the United States at the federal level have offered limited protections for older workers. Much the same can be said for Canadian laws at the federal and provincial level.

United States Federal Law

Federal legislation prohibiting age discrimination in the United States dates back to the 1960s, when Congress passed the *Age Discrimination in Employment Act of 1967* (ADEA), along with two 1964 laws barring discrimination against women and minorities: the *Equal Pay Act of 1963* and the Title VII of the *Civil Rights Act of 1964*; (Neumark 2003). Title VII prohibits employment discrimination based on an individual's race, sex, color, religion, or national origin, but omits age (Habinsky and Fitzgerald 2011).

The ADEA provides discrimination protection to workers, aged 40 and older, who work in companies with 20 or more employees. It protects job applicants and employees from discrimination on the basis of age in hiring, promotion, discharge, compensation, or terms, conditions or privileges of employment. The ADEA covers individuals, partnerships, labor organizations and employment agencies, and corporations that engage in an industry affecting interstate commerce. The Act also controls state and local governments. The ADEA is enforced by the Equal Employment Opportunity Commission (EEOC).

The Age Discrimination Act of 1975 prohibits discrimination on the basis of age in programs and activities receiving federal financial assistance. The Act, which applies to all ages, permits the use of certain age distinctions, and factors other than age, that meet the Act's requirements. The *Age Discrimination Act* is enforced by the Civil Rights Center.

The *Americans with Disabilities Act of 1990* protects disabled workers in companies with 15 or more employees. The law requires that an employer provide reasonable accommodation to an employee or job applicant with a disability, unless doing so would cause significant difficulty or expense for the employer. A reasonable accommodation is any change in the workplace (or in the ways things are usually done) to help a person with a disability apply for a job, perform the duties of a job, or enjoy the benefits and privileges of employment (EEOC n.d.).

Although the ADEA acknowledges ageism as a social problem that disproportionately affects employees 40 years of age or older, it does not address ageism as a human rights issue (Powell 2010). Rather, the ADEA is an employment law that addresses ageism from an economic standpoint. It can award back pay and future loss, but there is no recovery for emotional distress or punitive damages, as is the case with Title VII of the *Civil Rights Act of 1964.* There has been little movement to reframe age discrimination from an economic issue to a human rights issue in the United States (Powell 2010).

Canadian Law

The federally enacted *Canadian Charter of Rights and Freedoms 1982* ("the Charter") contains equality rights identified in section 15 (equal protection of the law and equal benefit under the law). These specifically include age as a protected ground of discrimination. However, the *Charter* only applies to laws, policies and practices by federal, provincial or local level governments and those departments and agencies directly accountable to them. In recent years, government has divested some of its traditional responsibility for providing services (such as housing and health care) to the private sector, a trend that leaves the *Charter* less relevant to older adults (Spencer 2005).

The federal government earlier enacted the *Canadian Human Rights Act of 1985* which protects Canadians against discrimination by federally regulated industries or service providers, including banks, airlines, as well as federal departments and agencies. The Canadian Human Rights Commission accepts harassment complaints based on 11 grounds, including age discrimination. The Canadian Human Rights Commission (2010) describes harassment as "unwelcome behaviour that demeans, humiliates, or embarrasses a person, and that a reasonable person should have known would be unwelcome".

In marked contrast to the *Canadian Charter of Rights and Freedoms*, provincial human rights laws apply to both public (government) and private bodies. All provinces and territories include age as a protected ground from discrimination in employment. However, it has been only in the last few years that forced retirement age caps have been removed in most of the jurisdictions that permitted compulsory retirement.

The province of Newfoundland and Labrador has included a special section on harassment in its new *Human Rights Act of 2010*. Even when other provincial human rights laws do not specifically refer to harassment, it is now *read into* the law and interpreted as a method of employment discrimination. However provincial human rights law in Canada identifies workplace harassment as a prohibited act only where it is tied to one of the specified protected characteristics such as age, race, sex, disability, or family status. Ontario uses this approach. New Brunswick, for example, describes harassment as any unwelcome comment or conduct that is directed toward an employee; *is based on grounds protected by human rights law;* is degrading, offensive or threatening; and is known or ought reasonably to be known to be unwelcome. Where harassment is not linked to a protected grounds or categories (sometimes called general harassment or personal harassment), the person's only option is to sue the employer for forcing him to resign because of intolerable work conditions (*Shah v. Xerox Canada Ltd.* 1998).

For the older worker, discrimination in employment is often not just about age. It can also reflect the person's marital status (e.g., widowed), family status, disabilities that the older worker may have, or that employers expect the older worker to develop. Older adults may seek non-discrimination assistance under disability provisions. For example, the Newfoundland and Labrador *Human Rights Act of 2010*, covers disability where the person has, or is believed to be developing, a disability. Human rights law in Canada, at least in theory, recognizes that people can experience discrimination by virtue of several intersecting categories. In adjudicative practice, however, the pressure is for claimants to "pick one" category that is most likely to result in a successful claim.

The Quebec *Charter of Human Rights and Freedoms* ("Quebec Charter") (the province's human rights law) is probably the most detailed provincial human rights law in the country. In addition to the typical non-discrimination clause in s. 10, the *Quebec Charter* offers workers of any age the "right to fair and reasonable conditions of employment with proper regard for the person's health, safety and well-being" (s. 46), as well as protections of "The right to life, personal security, inviolability and freedom, as well as the recognition of the legal personality of the individual" (s. 1).

One of the most noteworthy aspects of human rights law in Canada is the duty to accommodate:

> The duty to accommodate is the obligation to meaningfully incorporate diversity into the workplace. The duty to accommodate involves eliminating or changing rules, policies, practices and behaviours that discriminate against persons based on a group characteristic, such as race, national or ethnic origin, colour, religion, age, sex (including pregnancy), sexual orientation, marital status, family status and disability (Canadian Human Rights Commission 2005).

This duty exists for employment and other services, and covers all protected group characteristics. An employer's duty to accommodate extends to the point of undue hardship for the business. This places an important responsibility on employers to make significant adjustments to the workplace to meet the needs of workers in training, hours of work, working conditions, caregiving responsibilities and other needs. In addition, family status is a protected category in many Canadian human rights laws (Ontario Human Rights Commission n.d; Ontario Human Rights Commission 2009).

Older workers who provide care to a spouse, partner, grandchild, or other person, may face workplace harassment due to employers' beliefs that such caregiving interferes with workers' job responsibilities, in addition to the common lack of workplace accommodation for caregiving responsibilities that arise (Porter 2009–2010; Ontario Human Rights Commission 2002, 2006). The Ontario Human Rights Commission (2007) noted that employment decisions also should not be influenced by stereotypes about caregivers:

> Those who provide family care, or are perceived to, may wrongly be considered less competent, committed or ambitious than others – often due to gender stereotypes – and may be passed over for promotions, learning opportunities and recognition. Where workplace structures, policies, procedures or culture exclude or disadvantage persons with caregiving responsibilities, employers have a duty to consider adjustments to reflect such needs. This may involve, for example: providing flexible scheduling; permitting employees to take leaves of absence to care for family members who are aging, ill or have a disability or allowing alternative work arrangements.

The Ontario Human Rights Commission (2007) also noted that "creating a flexible and inclusive workplace benefits all employees, and advantages employers in hiring, retaining and getting the best possible performance from employees". The duty to accommodate under "family status" needs has not been adjudicated much to this point, but will become an important issue in the future, especially for older workers.

Barriers to Implementation of Age Discrimination Laws

In order for a law to benefit an older worker it must effectively prevent harms from occurring or offer meaningful and timely access to remedies. It must be responsive to older workers' needs and acknowledge their realities. For many reasons, at present, age discrimination law has a long way to go in protecting older workers.

In the United States, the procedural requirements for an ADEA lawsuit are complicated. Before an individual can sue in his/her own right, a private plaintiff must file charges with the EEOC or with an appropriate state agency. The EEOC may also sue to enforce the ADEA. There is a 3-year statute of limitations for both government and private lawsuits, starting from the date of an alleged willful violation. For cases of non-willful violations, the statute of limitations is 2 years from the date of the alleged violation.

There have always been several barriers to the use of human rights laws for older workers in Canada, including: lack of awareness of worker rights due to little public education; the complexity of the administrative law process once the worker files; length of time to decision (often years); the substantial risk of an adverse negative decision, especially for workers without legal representation; and law being interpreted adversely to the older worker's interests. Age discrimination cases have come infrequently to the attention of human rights tribunals in Canada and most have focused on mandatory retirement, rather than other aspects of employment.

Compared to 10 years ago, there are many more barriers to accessing the human rights system in Canada for all types of discrimination, including age discrimination. The last 10 years have witnessed the loss of the Court Challenges Program (which allowed lawyers to take major systemic issues all the way to the Supreme Court of Canada for interpretation), along with disappearance of Human Rights Commissions and Commission resources (Baker 2011; McNaughton 2011). In several jurisdictions, human rights systems have moved away from the human rights commission (which provided prevention, education, and mediation) to reactive "direct access" programs where cases go directly before tribunals, often without representation. Some Canadian jurisdictions, such as Saskatchewan, are contemplating to have cases directly heard by the court (Day 2010). That is an approach from the 1950s that was well recognized for its negative impact on individuals.

The ability of those who frame and interpret the issues to recognize, understand and appropriately respond to ageism is another factor in the efficacy of the law. Ageism is so tightly woven into our society that people scarcely notice it for what it is. The Ontario Human Rights Commission (2000) noted:

> Age cases tend to be treated differently than other discrimination cases, … The most noticeable difference from a human rights perspective is the lack of a sense of moral opprobrium linked to age discrimination which in comparable circumstances would generate outrage if the ground of discrimination were, say, race, sex or disability (pp. 3).

Employers, lawyers and other actors in the legal system have treated age discrimination as a lesser offense than race, sex or other discriminations, in part because it is a status we grow into; in other words, we all grow old (Ontario Human Rights Commission 2000; Charters 2008, pp. 5) There is no established theoretical framework for understanding and conceptualizing ageism and age discrimination, despite the efforts to create such a framework made by the Law Commission of Canada from 2009 to 2012 and the Law Commission of Ontario from 2009 to 2012.

To a large extent, much of the academic focus on age in human rights protections has centred on the minutiae of mandatory retirement (removing age caps or determining their justifiable boundaries, such as exceptions to age discrimination in insurance and pension benefits). The socially hidden parts of ageism in employment are a much larger iceberg of attitudes, stereotypes, and institutional practices. The ways in which these factors can create a chilling, non receptive environment for older workers have been largely ignored (Spencer 2003).

Courts have also been known to be ageist in their perspective on (or interpretation of) age discrimination cases. Supreme Court of Canada Chief Justice LaForest, in the seminal mandatory retirement case of *McKinney v. University of Guelph*, implied that requiring people to retire at a defined age was justifiable, because aging meant decrepitude. He went on to state at page 294–295: "There is a general relationship between advancing age and declining ability whereas no correlation exists between any of the other enumerated grounds and ability".

North American Laws against Workplace Harassment, Bullying and Abusive Work Environments

United States

Over the past 9 years there has been a sustained effort in the United States to legislate against harassment in the workplace. The impetus for this effort has reflected concerns about the growing extent of the problem in the workplace, and its negative effect on businesses, the people targeted, and other employees. Since 2003, 21

American states have introduced a version of the anti-bullying *Healthy Workplace Bill* (HWB) that would ban bullying in the workplace. The HWB is modelled on anti-discrimination law and is intended to "address the most egregious, harmful and severe forms. The HWB puts the misconduct on par with domestic violence and other potentially traumatizing experience" (Healthy Workplace Bill 2011).

California became the first U.S. state to consider the legislation. The California Healthy Workplace Bill would have amended the state's Fair Employment and Housing Law, which prohibits discrimination on the basis of enumerated characteristics such as age, gender and race (California Fair Employment and Housing Act, n. d.). The amendment would have made it an unlawful employment practice to "subject an employee to an abusive work environment." The bill defined an "abusive work environment" as "a workplace where an employee is subjected to abusive conduct that is so severe that it causes physical or psychological harm to the employee."

These healthy workplace laws shift much of the responsibility of "fixing the harassment problem" from the targeted employee to the employer. Healthy workplace laws also redress inadequacies in antidiscrimination laws (which focus on specific grounds or categories of discrimination or harassment, such as race, age, disability, and sex). In addition, healthy workplace laws displace the unreasonably high level of standard of conduct that is needed to be proven in private tort cases for intentional emotional distress, which required a threshold of "outrageous conduct". Getting a HWB law enacted in any state, however, has proven difficult. As of November 2011, there were 12 states considering 18 versions of anti-bullying bills, each at different stages of the legislative process. None was in force.

The HWB provides a legal remedy for employees who are subjected to an abusive work environment by allowing them to sue both their employer and the alleged bully for monetary damages. The HWB sets a high standard for misconduct and protects employers from liability (Habinsky and Fitzgerald 2011). If the employer has a policy in place (a preventive act) and enforces it (so the situation is corrected promptly), the employer avoids liability. That does not mean the work environment will be harassment free, just that there are steps in place to reduce and address its occurrence. In order for an individual to collect damages, the laws require the person to prove harm, which is shown by evidence from a licensed health professional that shows a negative health impact to the targeted worker.

Anti-bullying legislation imposes a form of "civility code" upon the workplace. It sets a high standard for misconduct; gives employers a reason to terminate or sanction offenders; and plugs gaps in current state and federal civil rights protections.

Canada

In Canada over the last decade, the courts have taken steps to recognize, at common law, the importance of a safe workplace. For example, in *Stamos v. Annuity Research & Marketing Service Ltd.* (2002), the Ontario Superior Court recognized that an employer owes a duty to its employees to provide a decent, civil and respectful workplace. An employee can sue for constructive dismissal if the person has to quit because of intolerable working conditions, such as that brought on by psychological harassment, and persistent and unwarranted criticisms (Chainey 2010). Efforts at legislating non-harassment in the workplace tend to come under either employment/ labour codes or occupational and health safety laws. In Canada, labour law rests with two levels of government. Labour law affecting most workers falls under provincial jurisdiction; the federal government has authority over federal public servants and a select number of specific industries, such as federal police services Canadian Human Rights Commission (March 2006), (i.e., the Royal Canadian Mounted Police), banking, and airlines.

Federal Law

There have been some federal efforts to develop anti-harassment policies for workplaces covered by the Canadian Human Rights Act (Canadian Human Rights Commission 2006). There have been very modest efforts over the past 8 years to enact new federal laws prohibiting harassment in the workplace. In 2010, before the House of Commons, a Member of Parliament tabled a private member's bill applicable to the federal jurisdiction, *Bill: C-276: "Workplace Psychological Harassment Prevention Act 2011"*.

Although it did not progress past the first reading, Bill C-276 would have established that every covered employee was entitled to employment free of psychological harassment. It required employers to establish anti-harassment policies and placed responsibility on other employees to inform their superiors about harassing behaviours in the workplace, meaning that they could no longer remain silent witnessess. It created a Committee to receive complaints and investigate psychological harassment of federally covered employees and report to the Public Service Commission. It offered mediation for hostile, inappropriate and unwanted conduct. The remedies for employees would have been similar to those in the Quebec law, discussed below. A number of remedies could have been ordered under Bill C-276. These may include reinstatement, payment of compensation, exemplary and moral damages, other action necessary to stop the harassment, medical and psychological support for the targeted worker, and training. The Bill imposed fines on employers of up to $10,000 for retaliation.

Since 2003, similar bills to add an anti-harassment law to the Canadian Labour Code have been tabled several times. These previous efforts include Bill C-360 in 2005 and Bill C-487 in 2007. However, private members' bills have little chance of being passed into legislation (Currie 2010). Although these bills appear to focus on prevention, education and early intervention, the targeted worker is still the focus, as these bills require the target to first let the harasser know this is unwanted behaviour.

Provincial Law

The first anti-bullying law in North America came into effect on June 1, 2004 when Quebec amended its *Labour Standards Act* to deal with psychological harassment in the workplace. It states that every employee has a right to a work environment free from psychological harassment. According to the Interpretation Guide for the *Labour Standards Act*:

> This responsibility falls on the employer and not on the person presumed responsible for the psychological harassment. It is the employer who has the responsibility of providing the employees with fair and reasonable conditions of employment and of respecting their health, safety, dignity and psychological and physical integrity (Commission des normes du travail 2011).

This amendment places responsibility on the employer to "take reasonable action to prevent psychological harassment and, whenever aware of the behaviour, to put a stop to it." It allows the employee to file a complaint with the *Commission des relations du travail* (the provincial labour relations board). It focuses on vexatious behaviour (behavior that is humiliating, offensive or abusive for the person on the receiving end that injures the person's self-esteem, causes anguish, and exceeds what a reasonable person would consider appropriate within the context of his work). In other words, it uses the "reasonable person" as the standard for gauging prohibited behaviour.

If the *Commission des relations du travail* considers that the employee has been the victim of psychological harassment and that the employer has failed to fulfill the imposed obligations, there are several possible remedies. These include ordering the reinstatement of the employee and indemnifying the employee for lost wages or for loss of employment. The employer can be ordered to take reasonable action to put a stop to the harassment; to pay punitive and moral damages to the employee; to pay for the psychological support needed by the employee for a reasonable period of time determined by the Commission; and to modify the employee's disciplinary record. Bill C-276 offered similar remedies.

The province of Saskatchewan made changes to the *Occupational Health and Safety Act* in 2007, which placed a responsibility on the employer to "ensure, insofar *as is reasonably practicable*, that the employer's workers are not exposed to harassment with respect to any matter or circumstance arising out of the workers' employment" (Occupational Health and Safety Act 1993, s, 3 (c)). It also places duties on employees not to harass or participate in the harassment of others.

Under the Saskatchewan legal definition, "harassment" means any inappropriate conduct, comment, display, action or gesture by a person that is based on one of the usual protected categories in human rights. (These categories are race, creed, religion, colour, sex, sexual orientation, marital status, family status, disability, age, nationality, ancestry or place of origin; they also include physical size or weight.) Harassment also includes behaviours that adversely affect the worker's psychological or physical well-being. Thus, the Saskatchewan provision covers what is commonly known as general or personal harassment. There are still important thresholds: the behaviour must be sufficiently grave that the harasser knows or ought reasonably to know it would cause a worker to be humiliated or intimidated, and the behaviour constitutes a threat to the health or safety of the worker. The Act provides no specific remedies.

In 2009 Ontario amended its *Occupational Health and Safety Act* (Bill 168) to address workplace violence and harassment. Under the amendments every employer must develop and post written policies on workplace violence and workplace harassment; conduct a risk assessment for workplace violence; develop a workplace violence and harassment program; and establish how incidents, complaints, or threats of workplace violence will be investigated and managed. Incidents or threats of workplace violence must be reported to the employer or supervisor. Employers have a responsibility to train employees in these policies and procedures. Employees can refuse to work where they have reason to believe that they are in danger of being a victim of workplace violence.

The BC Human Rights Coalition (2011) has been working with BullyFreeBC to highlight the need for anti harassment legislation in British Columbia. In consultation with its members BullyFreeBC has proposed changes to British Columbia's *Workers Compensation Act*. This would include amending the Act to include a section dealing with "Personal Harassment".

Legislative Efforts on the European Stage

While the primary focus of this chapter is on North America, it is instructive to contrast its legislative history and status of anti-bullying laws with those of the European Union (EU). The latter are more developed and may provide some direction for the future of similar legislative initiatives in North America (US and Canada).

The European Union (EU) has developed legislation on age discrimination that applies to its 27 member countries. Traditionally, the emphasis of European Union employment law centred on preventing sex discrimination or discrimination based on nationality. Since 1997, EU institutions have been granted new legal powers to combat discrimination on a number of grounds, including age, which has led to legislation against discrimination in employment. In 2000 there were at least two directives that further aided the process.

Employment Equality Directive 2000/78/EC established a *framework agreement* for equal treatment and occupation. Age was one of the included grounds. "Framework agreement" is the term used to describe the successful outcome of the European Social Dialogue. European Social Dialogue is a formal consensus process that brings together representatives from trade union and employer organizations to create a framework for discussions, negotiations and joint actions undertaken by the European Social Partners.

By 2006, nearly all EU member States had introduced the legal principles of this Council Directive into their national discrimination laws, including those pertaining to labour and employment law. The EU Employment Equality Directive (2000/78/EC) implements the principle of equal treatment in employment, training and membership in workers' or employers' organizations. It forbids direct and indirect discrimination (Article 2[1]); harassment; instructions to discriminate; and victimization; and stipulates avenues for legal redress and rules regarding the burden of proof (Article 10 [2]).

Comparing European and North American Laws on Harassment

In 2006 the EU developed a Framework Agreement on Harassment and Violence at Work (European Social Dialogue 2007). Under this agreement, each year, Member States identify the steps they have taken to implement the framework in their laws and policies. There are similarities and differences in the approaches being taken (European Social Partners 2011).

In Canada, actions are considered as harassment in the workplace if "the reasonable person" would have known the behaviour was unwelcome, not that the bully intended, knew or cared that it was harmful to the individual. This standard is an improvement for workers. However, the "reasonable person" can still be age-biased. A European Union Framework Agreement uses the term "harassment and violence" to describe workplace bullying. It defines these terms this way:

> Violence occurs when one or more workers or managers are repeatedly and deliberately abused, threatened and/or humiliated in circumstances related to work. Violence occurs when one or more workers or managers are assaulted in circumstances relating to work. Harassment and violence may be carried out by one or more managers or workers, with the purpose or effect of violating a manager's or worker's dignity, affecting his/her health and/or creating a hostile work environment.

Interpretation of actions such as sexual or racial harassment can be affected by the harasser's and target's gender, race, socioeconomic status and many other factors (Porter 2003–2004). Interpretation of harassment for an older worker may be affected similarly. The interpretation of workplace behaviour may reflect younger or middle aged people's view of what are "reasonable" comments and actions. These interpretations may ignore generational differences (for example, in determining what is acceptable language in the workplace). These generational differences in language may characterize the words used by both younger and older persons. Younger persons may use vulgar words not acceptable to older persons; older persons may use racist or sexist language that is not acceptable to younger persons. Researchers who study workplace harassment must include an age-specific analysis of what harassment in the workplace looks and feels like for older workers, and how they respond. Blackstone (2010), in a study of harassment among older workers, reported that harassment in the workplace is manifested in different ways for older and younger people. The most common response of older workers who were harassed in the workplace was to tell no one (Blackstone 2010). Younger workers who are harassed may respond differently. (For a detailed discussion of harassment of older adults in the workplace, see Chap. 3.)

Even workplace educators and trainers may inadvertently fall prey to ageist ideas: In Quebec, one example of "What is *not* harassment" offered in web-based training, described an employer introducing computers to the plant site to improve efficiency. However, the example provided was a situation involving a male employee "over 50" who expressed reluctance and resistance to be trained on a computer (Commission des normes du travail 2011). This example did little to illustrate what is not harassment, but it did feed the common stereotype of older workers as inadaptable and resistant to training. An example like this does not acknowledge that there might be legitimate reasons underlying the worker's "reluctance" (e.g., the educator's training style, or the worker's lower formal education or hidden literacy problems, factors that, from a human rights perspective, might need accommodation).

The process of enforcing workplace rights and the attendant costs are important issues in this area. The US *Healthy Workplace Bill* does not involve state agencies to enforce any provisions of the law. That has been used as a selling point to states in enacting and implementing the law. It is an economically efficient approach to cost containment, but it means the harassment law may lack teeth. The Bill also requires plaintiffs to use private attorneys, incurring costs to enforce their right to be free from harassment. This disadvantages workers who are poor.

Relying on occupational health and safety provisions to address workplace harassment has come under criticism. This is because the theoretical foundation of most occupational health and safety law is one of shared responsibility of employer and employee. Within that framework, the targeted older employee is expected "to do her part to fix" the problem of the harassment. This may make sense in terms of asking witnesses to report the harassment. However, placing the responsibility of fixing the problem on the target flies in the face of the dynamics of harassment, particularly, the power and control of the harasser. It treats harassment as an interpersonal rather than an organizational matter. Harassment, as an issue at law, is usually viewed as resulting from personal conflict (a one-to-one behaviour intended to target a specific individual based on specific characteristics) rather than from a deficient workplace environment or business culture (for a more detailed discussion of organizational culture, see Chap. 6). This view distracts attention from the consequences of a workplace environment that is unsupportive of older workers, is chilling toward the rights of older workers, or is characterized by multiple organizational actors.

In Ontario, new harassment provisions require the employer to have anti-harassment policies in place. The inspectors for the Ministry of Labour will not be able to take action against workplace harassment, other than to review whether an employer has developed policies and procedures for reporting and investigating harassment. Inspectors will not investigate specific complaints of harassment. They can only write orders if employers have not complied with the legislation. In other jurisdictions, there are also procedural issues that arise in the course of these laws that work against employees who pursue harassment complaints. For example, there is no appeal from the Quebec Commission decisions.

Although policies and grievance procedures have some effectiveness, what is really needed is a change in attitudes about how people interact in the workplace. No matter how good the wording is, no policy, contract provision or complaint procedure has meaning unless organizational leadership, at all levels, clearly and unequivocally opposes harassment, irrespective of its basis.

Conclusion

This review of current and pending legislation to protect older workers against age discrimination, harassment and other forms of workplace abuse reveals that progress has been made in North America and Europe. However, much remains to be done.

The laws discussed in this chapter largely view older workers as a homogenous group, without reference to important intra-group differences based on factors such as gender/gender roles and disability/health. In the future, these laws should consider how ageism intersects with other types of discrimination. In addition, employers, managers, legislators, judges and regulators should give greater attention to the definition of concepts such as discrimination, harassment and bullying. Finally, legislators should ensure that workplace laws and legal procedures allow older workers to make effective use of existing legal remedies. Until progress is made in these areas, older workers' situations will continue to be precarious.

References

Age Discrimination Act of 1975 42 USC. Sections 6101–6107

Age Discrimination in Employment Act of 1967, Pub. L. No. 90-202 Code, 29 U.S.C. § 621 through 29 U.S.C. § 634.

Americans with Disabilities Act of 1990, 42 USC Section 12101. Retrieved from http://www.ada.gov/pubs/adastatute08.htm. Accessed 1 Jan 2012.

Baker, D. (2011). http://www.workplaceviolence.ca/downloads/Ontario_Guide_Bullying_in_the_workplace.pdf Emerging issues in disability. In: *Proceedings from "Human rights in Canada: new realities, new directions" conference*. Canadian Association of Statutory Human Rights Agencies (CASHRA), Calgary, AB.

BC Human Rights Coalition (2011, May). Anti-bullying legislation in Canada. *News from the BC Human Rights Coalition, 11*(2), p. 1. Retrieved from http://www.bchrcoalition.org/files/documents/NewsletterMAY2011.pdf. Accessed 21 Mar 2012.

Blackstone, A. (2010). *Harassment of older adults in the workplace*. Presented at meeting of the American Sociological Association.

Braun v. Profile Seismic Ltd. 2009 AHRC 3. Human Rights Panels of Alberta (2009, June 16). *Diane Colley-Urquhart adjudicator*. Retrieved from http://www.canlii.org/en/ab/abhrc/doc/2009/2009ahrc3/2009ahrc3.html. Accessed 1 Jan 2012.

Butler, R. N. (1975). *Why survive? Being old in America*. New York: Harper & Row.

Cahill, K. (2011). Should older workers step aside? *Huffington Post*, August 28, 2011. Retrieved from http://www.huffingtonpost.com/kevin-e-cahill/should-older-workers-step b 941020.html. Accessed 4 Oct 2011.

California Healthy Workplace Bill (Assembly Bill 1582). California Legislature—2003–04 Regular Session.

California Fair Employment and Housing Act (n.d.). Government Code §§12900–12996.

Canada Safety Council (n.d.). *Bullying in the workplace*. Retrieved from http://canadasafetycouncil.org/workplace-safety/bullying-workplace. Accessed 1 Jan 2012.

Canadian Human Rights Act R.S.C., 1985, c H-6.

Canadian Charter of Rights and Freedoms Part I of the Constitution Act, 1982, 1982, c. 11 (U.K.), Schedule B.

Canadian Human Rights Commission (2005). *Duty to accommodate – frequently asked questions*. Retrieved from http://www.chrc-ccdp.ca/pdf/DTA_FAQ_en.pdf. Accessed 1 Jan 2012.

Canadian Human Rights Commission (2006, March). *Anti-harassment policies for the workplace: An employer's guide*. Retrieved from http://www.chrc-ccdp.ca/publications/anti_harassment_toc_eng.aspx#11. Accessed 11 Nov 2011.

Canadian Human Rights Commission (2010). *Human rights issues in 2010*. Retrieved from http://www.chrc-ccdp.ca/publications/ar_2010_ra/page4-eng.aspx#5. Accessed 21 Mar 2012.

Catalyst (2011). *Women's earnings and income*. Retrieved from http://www.catalyst.org/file/724/qt_womens_earnings_and_income.pdf. Accessed 27 Sep 2012.

Chainey, A. (2010). *Workplace violence and psychological harassment: An overview of the Quebec experience*. Paper prepared for the National Administrative Law, Labour & Employment Law and Privacy & Access Law Conference. 26–27 Nov 2010, Ottawa, ON.

Charter of Human Rights and Freedoms, R.S.Q., c. C-12.

Charters, S. (2008). *Age related policies: A global review on age discrimination legislation*. International Federation on Ageing. Retrieved from http://www.silverinnings.com/docs/Elder%20Law/International/Age%20Related%20Policies.pdf. Accessed 1 Jan 2012.

Civil Rights Act of 1964 PL. 88–352. (1964, July 2). 78 Stat. 241, as amended, Title VII.

Commission des normes du travail (2011). *Labour Standards Act. Interpretation guide, section 81.18*. Retrieved from http://www.cnt.gouv.qc.ca/en/interpretation-guide/part-i/act-respecting-labour-standards/labour-standards-sect-391-to-97/psychological-harassment-sect-8118-to-8120/index.html. Accessed 1 Jan 2012.

Currie, M. B. (2010, June). *Workplace harassment investigations under the Occupational Health and Safety Act*. Presented at the Ontario Bar Association/Law Society of Upper Canada conference "Workplace Violence and Harassment: Advising Clients on Amendments to Bill 168" held on March 10, 2010. Retrieved from http://www.greggowe.com/content/workplace-harassment-investigations-under-ontario-occupational-health-and-safety act. Accessed 11 Nov 2011.

Day, S. (2010, July 28). *Canada's human rights institutions at risk Women's Court of Canada*. Retrieved from http://womenscourt.ca/2010/07/canada%E2%80%99s-human-rights-institutions-at-risk. Accessed 1 Jan 2012.

Drolet, M. (2011, Spring). Why has the gender wage gap narrowed? *Perspectives on Labour and Income*. Statistics Canada Retrieved from http://www.statcan.gc.ca/pub/75-001-x/2011001/pdf/11394-eng.pdf. Accessed 1 Jan 2012.

Equal Employment Opportunity Commission (n.d.). *Charge statistics FY 1997 through FY 2010*. Retrieved from http:www.eeoc.gov/eeoc/statistics/enforcement/charges.cfm. Accessed 22 Nov 2011.

Equal Pay Act of 1963 Pub. L. No. 88-38, 77 Stat. 56 (codified as amended at29 U.S.C. § 206(d)).

European Union Framework Agreement on Harassment and Violence at Work. ETUC/BUSINESSEUROPE-UEAPME/CEEP, 2011.

European Social Partners (2011, October 27). Implementation of the European autonomous framework agreement on harassment and violence at work. Final joint report. Retrieved from: http://www.etuc.org/IMG/pdf/BROCHURE_harassment7_2_.pdf. Accessed 4 Oct 2012.

European Social Dialogue. (2007, April). Framework agreement on harassment and violence at work. Retrieved from : Accessed 4 Oct 2012.

European Union Employment Equality Directive 2000/78/EC (2000, November 27). Retrieved from : http://europa.eu/legislation_summaries/employment_and_social_policy/employment_rights_and_work_organisation/c10823_en.htm. Accessed 4 Oct 2012.

Habinsky, J., & Fitzgerald, C. M. (2011). Office bully takes one on the nose: Developing law on workplace abuse. *New York Law Journal*, January 21, 2011. Retrieved from http://workplaceviolencenews.com/2011/01/21/office-bully-takes-one-on-the-nose-developing-law-on-workplace-abuse/. Accessed 1 Jan 2012.

Healthy Workplace Bill (2011). *Why the US needs, and we are advocates for, the Healthy Workplace Bill*. Retrieved from http://healthyworkplacebill.org/blog/rationale. Accessed 11 Nov 2011.

Human Resources and Skills Development Canada (n.d.). *Indicators of wellbeing in Canada. Work-employment rate*. Retrieved from http://www4.hrsdc.gc.ca/.3ndic.1t.4r@-eng.jsp?iid=13. Accessed 9 Nov 2011.

Human Rights Act (2010). S.N.L. 2010, c. H-13.1.

Institute for Women's Policy Research (2012, March). *The gender wage gap: 2011*, IWPR Fact Sheet #C350. Retrieved from http://www.iwpr.org/. Accessed 21 March 2012.

Labour Standards Act R.S.Q., c. N-1.1 Division V.2 Psychological Harassment, s. 81.18 to 81.20 (s. 81.19).

Law Commission of Canada (2004) Does age matter? Law and relationshps between generations. Retrieved from: http://epe.lac-bac.gc.ca/100/206/301/law_commission_of_canada-ef/2006–10–20/www.lcc.gc.ca/pdf/does_age_matter.pdf. Accessed 4 Oct 2012.

Law Commission of Ontario (2012, April) A framework for the law as it affects older adults: advancing substantive equality for older persons through law, policy and practice. Retrieved from : http://www.lco-cdo.org/en/older-adults-final-report-framework. Accessed 4 Oct 2012.

Lowenstein, A., & Doron, I. (2008). Intergenerational justice and elder abuse. Intergenerational justice: An Israeli Perspective. *Indian Journal on Ageing*, *22*(3), 356–372.

Lutgen-Sandvik, P., Namie, G., & Namie, R. (2009). Workplace bullying: Causes, consequences, and corrections. In P. Lutgen-Sandvik & B. D. Sypher (Eds.), *Destructive organizational communication*. New York: Routledge Press.

McKinney v. University of Guelph (1990). 3 S.C.R. 299 at 294–295.

McNaughton, H. (2011, June). *Human rights systems*. In: Proceedings from "Human rights in Canada: New realities, new directions" Conference. Canadian Association of Statutory Human Rights Agencies (CASHRA), Calgary, AB Retrieved from http://cashra2011.ca/documents/2B-HumanRightsSystems-speakingnotesHMacnaughtonADD.pdf. Accessed 1 Jan 2012.

Namie, G. (2003). Workplace bullying: Escalated incivility. *Ivey Business Journal*, November/December.

Neumark, D. (July 2003). Age discrimination legislation in the United States. *Contemporary Economic Policy, 21*(3), 297–317.

Occupational Health and Safety Act (1993). S.S. c. O-1.1

O'Connell, M. (2005). *The role of training in preventing the labour market exclusion of older workers*. Centre for Public Policy, University of Melbourne. Retrieved from http://public-policy.unimelb.edu.au/__data/assets/pdf_file/0007/509452/Oco1.pdf. Accessed 1 Jan 2012.

Ontario Human Rights Commission (n.d.). *Fact Sheet #3: Employment discrimination based on family status*. Retrieved from http://www.ohrc.on.ca/en/resources/factsheets/famstat3. Accessed 1 Jan 2012.

Ontario Human Rights Commission (2000). *Discrimination and age – human rights facing older persons in Ontario: discussion paper* (2000, May 31). Retrieved from http:www.ohrc.on.ca/english/consultations/age-discussion-paper.pdf. Accessed 11 Nov 2011.

Ontario Human Rights Commission (2002). *Policy on discrimination against older people because of age*. Retrieved from http://www.ohrc.on.ca/en/resources/Policies/agepolicyen/pdf. Accessed 1 Jan 2012.

Ontario Human Rights Commission (2006). *The cost of caring: Report on the consultation on discrimination on the basis of family status*. Retrieved from http://www.ohrc.on.ca/en/resources/discussion_consultation/famconsult/pdf. Accessed 21 Mar 2012.

Ontario Human Rights Commission (2007). *Policy on discrimination against older people because of age*. Retrieved from http:www.ohrc.on.ca/en/resources/Policies/agepolicyen/pdf. Accessed 11 Nov 2011.

Ontario Human Rights Commission (2008). *Understanding ageism and age discrimination*. Retrieved from http://www.ohrc.on.ca/en/issues/age/age_discrimination_brochure_en.pdf. Accessed 21 March 2012.

Ontario Human Rights Commission (2009). *Family status*. Retrieved from http://www.ohrc.on.ca/en/issues/family/family_status_brochure_en.pdf. Accessed 30 Nov 2011.

Ontario Safety Association for Community and Health Care (2009). *Bullying in the workplace*. Retrieved from http://www.workplaceviolence.ca/downloads/Ontario_Guide_Bullying_in_the_workplace.pdf. Accessed 21 Mar 2012.

Pittman, K. (2010, November 18). *The plight of the older worker*. Retrieved from http://kvalleobs.blogspot.com/2010/11/plight-of-older-worker.html. Accessed 9 Nov 2011.

Porter, N.B. (2003–2004). Sex plus age discrimination: Protecting older women workers. Denver. *University Law Review, 81*, 79–110.

Porter, N.B. (2009–2010). Synergistic solutions: an integrated approach to solving the caregiver conundrum for "real" workers. *Stetson L Rev, 39*, 777–860.

Powell, M. (2010). Ageism and abuse in the workplace: A new frontier. *Journal of Gerontological Social Work, 53*(7), 654–665.

Schneider, J. A. (2011). *Who are the long term unemployed in this recession and what can be done to help them?* Policy report, George Washington University. Retrieved from http://www.thecyberhood.net/documents/papers/unemployment.pdf

Schwartz, K. (2010). *Unemployment among older adults stuck near record high*. National Council on Aging, October 12, 2010. Retrieved from http://www.ncoa.org/press-room/press-release/unemployment-among-older.html. Accessed 29 Nov 2011.

Shah v. Xerox Canada Ltd. (1998). O.J. No. 4349 (affirmed by the Court of Appeal [2000] O.J. No. 849]).

Spencer, C. (2003). *Falling off the retirement conveyer belt: Boomer women and mandatory retirement*. Canadian Association on Gerontology.

Spencer, C. (2005, October 29–30). *Why the Canadian Charter of Rights and Freedoms is becoming increasingly irrelevant to older adults*. Paper presented at the Canadian Elder Law Conference: Law, Community and Care, 2005, Vancouver, BC, Canada.

Stamos v. Annuity Research & Marketing Service Ltd. 2002-05-13 Ontario Supreme Court.

Statistics Canada (2011). Seniors internet use. *The Daily*, October 12, 2011, Retrieved from http://www.statcan.gc.ca/daily-quotidien/111012/dq111012a-eng.htm. Accessed 11 Nov 2011.

Szinovacz, M. E., & Davey, A. (2005). Predictors of perceptions of involuntary retirement. *Gerontologist, 45*(1), 36–47.

Workers Compensation Act. [RSBC 1996] c. 492.

Workplace Bullying Institute (2010a). *Age & workplace bullying: 2010 WBI Survey*. Retrieved from http://www.workplacebullying.org/multi/pdf/age_2010_wbi.pdf. Accessed 1 Jan 2012.

Workplace Bullying Institute (2010b). *Recession and workplace bullying: A 2010 WBI survey*. Retrieved from http://www.workplacebullying.org/multi/pdf/recession 2010 wbi.pdf. Accessed 11 Nov 2011.

Workplace Psychological Harassment Prevention Act (2011). Bill C-276, 40th Parliament, 3rd Session. Retrieved from http://www.parl.gc.ca/HousePublications/Publication.aspx?Language=E&Mode=1&DocId=4330022. Accessed 1 Jan 2012.

Chapter 10
International Protection for the Human Rights of Older People: History and Future Prospects

Israel Doron, Bethany Brown, and Susan B. Somers

In this chapter the authors seek to give gerontologists, lawyers and other interested practitioners in North America a better understanding of the global picture of ageism in the workplace as it relates to international and regional human rights agreements. The legal landscape regarding age discrimination is very different around the world. In many undeveloped countries "age discrimination" is unrecognized under local laws. Even in many developed countries, age-based employment policies, such as mandatory retirement, a maximum age limit for recruitment, and age-based workforce layoffs or redundancies are common and legal practices.

Today this field of law is in a state of flux. New laws, court rulings, and international instruments and policies are being established (Kye 2009; Field and Sappideen 2009). In 2011, an ongoing working group under the auspices of the United Nations (UN) began to study older people's human rights. These are times of change and this chapter will map and describe the progression of thinking around older people's rights from the perspective of the need for an international and binding convention in this field.

Historically, "old age" or "age discrimination" has been unknown or ignored under international human rights laws (Rodriguez-Pinzon and Martin 2003; Doron 2005). For example, when the UN's Universal Declaration of Human Rights

I. Doron, J.D. (✉)
Department of Gerontology, University of Haifa,
Mount Carmel, Haifa 31905, Israel
e-mail: idoron@univ.haifa.ac.il

B. Brown, J.D.
HelpAge International,
London, UK
e-mail: Bethanybrown.blb@gmail.com; bethanypurple@gmail.com

S.B. Somers, J.D.
International Network for the Prevention of Elder Abuse, Vancouver, Canada
e-mail: Sbsomers5@aol.com

P. Brownell and J.J. Kelly (eds.), *Ageism and Mistreatment of Older Workers: Current Reality, Future Solutions*, DOI 10.1007/978-94-007-5521-5_10,

(Universal Declaration), the general international commitment for equality in rights and freedoms, was established in 1948, it did not include "age" as a legally-prohibited distinction (Doron and Mewhinney 2007). Article 2 of the Universal Declaration states that everyone is entitled to the rights and freedoms set forth in the Declaration without any distinction, such as race, color, sex, language, religion, political or other opinion, social or national origin property, birth or other status. Here, age may be implied but is not specifically stated.

Moreover, the Declaration included very few references to older adults as a distinct category of people needing human rights protection An example is in article 5, which provides for the right to a standard of living that is adequate for the wellbeing and health of persons and their families, and details life events relevant to this right as including unemployment, illness, disability, widowhood, old age, or other lack of livelihood beyond a person's control.

The UN International Bill of Human Rights encompasses the Universal Declaration, the International Covenant on Civil and Political Rights (ICCPR) (1966), and the International Covenant on Economic, Social and Cultural Rights (ICESCR) (1966). This important collection of documents does not contain references to older people and they all reflect a lack of awareness and political ignorance of age discrimination and ageism, as for example Article 26 of the ICCRP, and Article 7 of the ICESCR.

Keeping the historical context in mind in the first section of this chapter, the authors discuss the development of the existing international human rights framework in light of the UN system of aging policy documents. Each system has benefits and limitations, with major distinctions between binding and non-binding mechanisms for enforcement. The second section of this chapter reviews three parts of the current landscape: the problems older workers face around the world; currently-unfolding developments at the UN on older people's rights; and the potential for an international convention on the human rights of older people. Such an agreement could be a game-changer for older workers in North America and around the world.

The Existing International Framework

Binding Versus Non-Binding Instruments

What does it mean to be *legally binding*? In international public law, the term legally binding represents a complex concept and raises questions such as: binding *to whom?* (individuals or states); binding *by what?* (by custom or by agreement); and binding *by whom* (The UN or states themselves)? For purposes of this chapter, the authors review public international law, which is based on contractual provisions, meaning international ***conventions***. Countries sign on, agreeing to be bound by their word to uphold an agreement or convention. To hold them accountable, there is an enforcement mechanism that is stated within the convention itself. This may take the form, for example, of a mechanism whereby complaints can be lodged and investigated. The extent to which a country is bound by international agreements is for every member state to decide for itself, regardless of its wealth or political status. Once it has agreed to a convention, if it does not uphold its obligations, and the rights conferred to the errant country's people are violated, those people can turn to the international community for redress through the enforcement mechanism.

Alongside the binding body of international law, there is also a non-binding element known as "soft law". In a complicated, divided and interest-driven international community, it is easier for nations to agree on ***declarations,*** which are non-binding in character. Despite the non-binding nature of declarations, they do play an important role. For example, although declarations are not formally binding, they have played an important role in the making of binding international law, commonly referred to as "hard law" (Gold 1983; Boyle 1999). The flexible nature of declarations can make it easier for states to implement them while taking into account their unique cultural and legal considerations—an option which does not always exist in "hard law" and binding instruments.

The authors will discuss non-binding and binding law at the international level, and then provide an overview of regional systems' treatment of older peoples' rights. Regional human rights systems around the world include the Organization of American States, the African Union, and the Council of Europe. The Arab League Council of the League of Arab States has published a charter of human rights (Arab Charter on Human Rights 1994).

Soft Law and Older Persons' Rights

Since the 1970s, a wealth of international soft law documents have been adopted by UN member states in the field of older persons' rights. This significant body of international declarations, proclamations, and plans of actions share one legal characteristic: they are not legally binding and they cannot be internationally or nationally enforced. They do not use the language of rights at all, but nonetheless show support and attention to the situations of older people.

This process began almost 40 years ago, as part of the 1974 UN World Population Conference, which adopted the World Population Plan of Action.[1] A crucial part of this plan of action was the recognition that the world is aging and there is "an urgent need...for the development of social security and health programs for the elderly."[2] As an outcome of this plan of action, the UN general assembly adopted a series of resolutions that culminated in a decision to convene, for the first time, a World Assembly on Ageing.[3]

---

[1] World Population Plan of Action 1974, in Rights of Older Persons, at 27.

[2] Ibid, Article 12.

[3] See Resolutions 33/52 of 14 December 1978; 34/153 of 17 December 1979, and 36/20 of 1981 – in Rights of Older Persons, *supra* note 3, at 31–33.

In 1982 the World Assembly on Ageing created the Vienna International Plan of Action on Ageing (VIPAA) which was endorsed by the UN General Assembly (United Nations 1998). This Plan of Action included significant declarations and statements. For example, it reaffirmed in the Preamble the belief that "the fundamental and inalienable rights enshrined in the Universal Declaration of Human Rights apply fully and undiminshedly to the aging" (p. 3). Moreover, it declared that all countries will "develop and apply at the international, regional, and national levels, policies designed to enhance the lives of the aging as individuals and to allow them to enjoy in mind and in body, fully and freely, their advancing years in peace, health and security". VIPAA concentrated on two facets of population ageing: humanitarian (responding to the specific needs of older persons); and macroeconomic or demographic (dealing with the implications of an ageing population for socio-economic policy). The latter focused on general concepts meant to ensure that older persons did not become a drain on national resources (UN Department of Economic and Social Affairs 2008).

As part of the implementation of VIPAA and marking its tenth anniversary, the UN General Assembly (GA) adopted the UN Principles for Older Persons (United Nations 1998), which correlate closely with the rights recognized in the ICESCR.[4] The Principles include independence; participation; care; self-fulfillment and dignity. This document was followed by another UN General Assembly (GA) Resolution, known as the Proclamation on Ageing. Among other things, the Proclamation established the International Day of Older Persons, and declared 1999 as the International Year of Older Persons (United Nations 1998).

The final important stage in the development of "soft-law" instruments in the field of rights of older persons came about in 2002, during the second World Assembly on Ageing in Madrid. Similar to the Vienna Assembly, this meeting concluded with the creation of the Madrid International Plan of Action on Ageing (MIPAA) (United Nations 2003). MIPAA is the product of the convening of over 156 countries by the then-UN Secretary General, Kofi Annan. Unlike VIPAA, which included more general declarations on humanitarian interventions and social welfare, MIPAA adopts a life-course, inter-generational, and developmental approach. It is more concrete, by setting both "priority directions" as well as specific objectives. The general spirit of the document is toward promotion of the participation of older persons as citizens with full rights. It seeks to assure that persons everywhere are able to age with security and dignity.

For example, under "Priority Direction 1: Older Persons and Development", MIPAA specifies an "Issue" on the topic of "Work and the Ageing Labour Force." Its objective is "Employment opportunities for all older persons who want to work." Toward that end, it recommends that countries should enable older persons to continue working as long as they want to work and are able to do so. MIPAA is

[4] ECOSOC, Comm. on Econ., Soc. & Cultural Rts., 13th session, GENERAL COMMENT NO. 6: THE ECONOMIC, SOCIAL AND CULTURAL RIGHTS OF OLDER PERSONS, Para. 5, U.N. Doc. E/C.12/1995/16/Rev.1 (1995).

a comprehensive document that touches upon almost all key areas relevant to older persons, not only identifying these areas but also setting priorities and objectives. Its major drawback is that it is a non-binding document; it lacks both targets and implementation monitoring.

MIPAA calls for the full employment of older persons in satisfying and productive work, linking it to continuing access to education and training programs (United Nations 2003). MIPAA builds on the conquest of world poverty, social integration, and full employment agreed upon in the Copenhagen Declaration and Programme of Action at the 1995 World Summit for Social Development. There, the world committed to the goal of full employment, and this commitment should be inclusive of older persons as long as they are able and willing to work, in the same spirit as MIPAA (Venne, n.d.).

In 2012, the world will mark the tenth anniversary of MIPAA. There is little reason for celebration. Because MIPAA (and the rest of the soft law instruments in the field) is not a legally binding document, there is no mandatory procedure through which member states regularly inform the UN about their implementation efforts. Such reports are scarce, and presently there is no comprehensive information about, or assessment of, national implementation or success in bettering the lives of older persons. Some recent UN reports regarding the follow-up to MIPAA suggest there are significant barriers to the successful implementation of MIPAA (e.g., lack of data, financial constraints, gaps between formal policies and realities and more).[5]

From a normative perspective, the authors conclude that the needs of older people have not been met by the Vienna Plan, the Madrid Plan, or the UN Principles for Older Persons. In these plans, created over the course of nearly two decades, nations have set policies that have been ignored, and action plans that have been abandoned. A statement of principles is not enough to create accountability on the part of states. There is no explicit reference to international human rights mechanisms. Furthermore, the existing discussion has focused on economic, social and cultural rights, not civil and political rights. Hence, it was not surprising that the Bonn 2009 UN Expert Group, which studied the rights of older persons on the international level, concluded:

> The UN policy documents on ageing and older persons provide a comprehensive treatment of issues and public policy strategies but as "soft law", they contain no legally binding obligations and there are no sanctions for non-performance. Instead, they incorporate norms and precepts, which governments agree to be guided by, but without any requirement to account for adherence. (p. 15) (Doron and Apter 2010b).

Binding Instruments

When moving from the soft law or non-binding sphere into the binding sphere, the picture is different. In contrast to the wealth of soft-law documents, there are almost no binding international documents dedicated or targeted to the rights of older

[5] For example, see Follow-up to the Second World Assembly on Ageing , 19 July 2006.

persons. No comprehensive and binding framework on the human rights of older people currently exists, and to date, international bodies have given only piecemeal attention to this issue. This reality has been mentioned in recent years by various scholars (Doron 2005; Megret 2011), and was also summarized by Rodriguez-Pinzon and Martin (2003):

> A strategy to have a comprehensive legal instrument on elderly rights is missing at the international level in both universal and regional systems. There are very few provisions in international law that directly address elderly rights. There are isolated efforts by certain international bodies to systematically refer to the rights of the elderly when interpreting their corresponding conventions....
>
> However, there is no specific international body with the mandate to focus on the rights of the elderly. Nor is there an elderly rights convention in place. It is in fact the only vulnerable population that does not have a comprehensive and/or binding international instrument addressing their rights specifically (Rodriguez-Pinzon & Martin, p. 1008) .

The best known human rights agreements—the International Covenant of Civil and Political Rights[6] and the International Covenant on Economic, Social and Cultural Rights,[7] —contain almost no explicit or direct reference to older persons. The most important development in securing the human rights of older people may have been made by the Committee on Economic, Social and Cultural Rights, in publishing its General Comment 6 on The Economic, Social and Cultural Rights of Older Persons.[8] This General Comment interprets and constructs the different relevant articles of ICESCR as they apply specifically to older persons. Tangentially, General Comment 19, on the Right to Social Security, clearly articulates these rights for older people.

The Convention on the Elimination of Discrimination against Women (CEDAW) has provided some of the strongest language for the human rights of older women in recent years in its General Recommendation 27. It highlights that women are less likely to be included in formal employment, and are often paid less for the same work as men while accruing fewer pension benefits for time spent out of the labor force. Older women in the labor force, the General Recommendation 27 notes, are often the sole caregivers for dependent children, spouses and relatives.[9] General Recommendation 27 provides policy guidance for protecting the human rights of older women in the workforce; 187 parties have signed on, making it one of the most universal of all recommendations.

Other documents focus on issues relating to older people in topical areas: the International Convention on the Protection of the Rights of All Migrant Workers and Members of Their Families (1990) mentions older people. The International Labour Organization has a Convention on Social Security in old age.[10] The fourth Geneva Convention, in its treatment of attending to civilians in armed conflict,

[6] In Rights of Older Persons, at 447.

[7] Ibid, at 449.

[8] Ibid, at 67.

[9] Committee for the Elimination of All Forms of Discrimination against Women, General Recommendation 27, para. 20.

[10] Ibid, at 337.

mentions older people in Articles 14 and 17.[11] These and other international documents explicitly mention older people, but only in provisions that group together a large catalogue of vulnerable persons. Thus, they have little impact and do not address the complexity of the rights of older persons.

Recently, the countries agreed to a Convention on the Rights of Persons with Disabilities at the UN. This most recent human rights convention includes key binding features that are of great importance and relevance for the rights of older persons (e.g., those involving legal capacity, guardianship, housing and accessibility). Some scholars have already pointed to the relevance of this convention to the advancement of rights of older persons (Kanter 2009). However, there is not enough experience with this convention to assess its implication for older persons.

The existing international system for securing the human rights of older persons is scattered. A mention here, and a nod there, have not created a unified system of claimable rights. A viable recognition of the rights of older people would be unified; it would pull together and clarify disparate rights. It requires a convention that interprets the rights of older persons through the already-agreed-upon language of human rights. Such a convention would empower older people from all walks of life, from those in nursing homes, to those pursuing their careers in the community.

Regional Instruments

In some world regions, the human rights of older people are reflected in various declarations and in case law established by regional courts. For example, the African Charter on Human and Peoples' Rights has almost no direct reference to older persons. However, its Article 29 stipulates that each individual must respect and maintain his or her parents in case of need (African Union 1986). Moreover, Article 18 states that "the aged and the disabled shall also have the rights for special measures of protection in keeping with their physical or moral needs." The African Commission on Human and Peoples' Rights are moving forward to clarify these provisions of the African Charter on Human and Peoples' Rights.

In the Inter-American System, consisting of North America, South America, Central America and the Caribbean, the current legal picture is not significantly different. The original binding documents on human rights, The American Declaration on the Rights and Duties of Man, promulgated in 1948 and the later American Convention on Human Rights of 1969, did not include specific reference to older persons. However, in 1988, the Conference of American States added the following provisions as part of the Additional Protocol to the American Convention on Human Rights (also known as the Protocol of San Salvador)[12]:

---

[11] Ibid, at 445.

[12] Additional Protocol to the American Convention on Human Rights in the Area of Economic, Social, and Cultural Rights, Nov. 17, 1988, art. 17. (Protocol of San Salvador),

Article 17: Protection of the Elderly: Everyone has the right to special protection in old age. With this in view the States Parties agree to take progressively the necessary steps to make this right a reality and, particularly, to:

(a) Provide suitable facilities, as well as food and specialized medical care, for elderly individuals who lack them and are unable to provide them for themselves;
(b) Undertake work programs specifically designed to give the elderly the opportunity to engage in a productive activity suited to their abilities and consistent with their vocations or desires;
(c) Foster the establishment of social organizations aimed at improving the quality of life for the elderly".[13]

This binding commitment was also supported by soft-law instruments such as the Declaration of the Latin American and Caribbean Meeting of Leaders of Organizations of Older People in 2001 (also known as The Declaration of Lima)[14]; and the Declaration of the Pan American Symposium on Ageing and Health: Innovative Approaches to Health and Social Services for the Elderly.[15]

As with the international UN, the regional Organization of American States is in the process of establishing a working group on the human rights of older persons. As early as 2012, it will produce a report on the human rights situation of older people in the Americas, as well as a draft Inter-American Convention on the Human Rights of Older Persons.[16]

The most sophisticated and strongest regional legal infrastructure is found in Europe. European agreements combine a traditional international convention (the European Convention for the Protection of Human Rights and Fundamental Freedoms 1950) with a novel supranational Charter of Fundamental Rights of the European Union (2000). Yet even the European Convention for the Protection of Human Rights does not include "age" among its grounds of prohibited discrimination listed in Article 14 "Prohibition of Discrimination."[17] Nevertheless, as a binding legal instrument, with enforcement mechanisms that allow for individual complaints to an international court, there is ample evidence of older persons exercising their rights by law under this convention. Indeed, the European court of human rights has heard many cases regarding social security for older people after retirement and has upheld the rights of older people to fair trials, to property (in their pensions), and, potentially, to freedom from discrimination on the basis of an "other status" (Rodriguez-Pinzon and Martin 2003). The European experience proves that when an enforceable convention exists, older persons can take advantage of it, and promote their rights through an international mechanism.

[13] See Protocol of San Salvador, Article17.

[14] See in Rights of Older Persons, supra note 1, at 174.

[15] Ibid, at 180.

[16] OEA/Ser.G CP/CAJP-3017/11, Installation of the Working Group Envisaged Under Resolution AG/RES. 2654 (XLI-O/11), 14 September 2011.

[17] Ibid at 239. It should be noted however that in Article 21 of the European Charter of Fundamental Rights, regarding "Non-discrimination" age as a specific legal category has been included.

The Islamic world has also established a soft-law document on the rights of older persons. As part of a symposium held in Kuwait during 1999, the Islamic countries adopted a declaration known as the Declaration on the Rights of Elderly – an Islamic Perspective.[18] This declaration included statements such as, "The elderly have rights that should be recognized and admitted by their communities" (Article 3). However, this Declaration has no binding force.[19]

These various documents represent advances over an earlier era when no protection existed for the human rights of older persons. However, the overall picture is similar to that in the general international sphere: On the regional level there is a large collection of "soft law" declarations along with very few, scattered and minimally binding "hard" law conventions that address the human rights of older persons. In most cases, references to older persons are presented in the context of dependence and weakness. And there is little evidence that provisions to safeguard the human rights of elderly and dependent persons are used in international tribunals. The only exception is in Europe, where both the European Court of Human Rights and the European Court of Justice have been instrumental in promoting rights of older persons and enforcing European international and supra-national hard-law instruments.

Current State of the World's Older Workers

For many older people, livelihoods depend on their ability to maintain employment. Older people rank among the poorest people in the poorest countries around the world (Barrientos et al. 2003). The rights that older people in the US take for granted are not even thought of as rights in many parts of the world. For example, anti-age discrimination laws in employment often do not exist or are perceived as novel. Retirement decisions are dictated by mandatory retirement ages, which are still legal and prevalent in most countries in the world (Office of the High Commission for Human Rights 2010). Less than 20 % of the world's population receives any sort of pension (UN Department of Economic and Social Affairs 2007).

The right to work free of abuse in the workplace is a civil right. Article 1 of the ICCPR declares "All peoples have the right of self-determination. By virtue of that right they freely determine their political status and freely pursue their economic, social and cultural development." All older people have a right to employment and pursuit of economic development with dignity and freedom from discrimination and abuse in the workplace.

Implementation of an international convention on the human rights of older people would have real and immediate gains for older workers. By outlawing age

[18] See in Rights of Older Persons, supra note 1, at 234.

[19] See also The Arab Plan of Action on Ageing to the Year 2012 (APAA). Online (last accessed: October 3, 2011) http://www.escwa.un.org/popin/docs/ArabPlanofActionAgeing2012Eng.pdf

discrimination in hiring practices, older people would be free to apply for and retain positions for which they are qualified. Today this is not the case in much of the world. For example, it is not uncommon for employment ads from other countries to state that applicants must be under the age of 30. The Supreme Court of Bangladesh recently ousted the Nobel Prize-winning Muhammad Yunus from his post as managing director of Grameen Bank, under a retirement law requiring that older workers obtain government approval to remain in the workforce after retirement age (Pilapil 2011).

In addition to age, health is a key factor in the ability of older workers to remain in the workforce. In a 2011 report, UN Special Rapporteur on the Human Right to Health, Anand Grover, noted, "Society should move beyond seeking simply healthy ageing for its citizens, and begin working towards active and dignified ageing, which should be planned and supported just like any other stage of the individual's life course" (Human Rights Council 2011).

Current and Future Developments

UN's Open-Ended Working Group on Ageing

In 2010, as a follow-up to the 2002 Second World Assembly on Ageing, the UN General Assembly established an Open-Ended Working Group "for the purpose of strengthening the protection of the human rights of older persons by considering the existing international framework of the human rights of older persons and identifying possible gaps and how best to address them, including by considering, as appropriate, the feasibility of further instruments and measures…"[20]

At its first meeting in April 2011, the Working Group explored gaps in the existing legal framework. It heard from representatives of member states, the International Labour Organization, the United Nations Population Fund, the Office of the High Commissioner for Human Rights, and representatives of NGO's.

At its next meeting in August 2011, the discussion continued and became more specific. The European Union and its member states came out staunchly against a convention while many Latin American countries stood strongly in favor. Few Asian or African countries were present. In order to gain true accord, full participation of all regions is necessary for the legitimacy of any agreement and the true representation of older people. Developing countries with young populations should remember that life expectancy at age 65 does not differ greatly between the developed and the developing world (UN Department of Economic and Social Affairs 2009). Older people in developing countries are particularly invisible, yet have a necessary and unique role to play in their countries' development. Thus, the authors hope that Asian and African countries will

[20] UN Document A/RES/65/182 Follow-Up to the Second World Assembly on Ageing.

be better represented at the next meeting of the Open-Ended Working Group, which will be in the summer of 2012.

A Convention

The need for a specific and unique convention for the rights of older persons is on the frontlines of international debate. Several commentators have made strong arguments in favor of such a convention. For example, Tang and Lee (2006) argued that:

> Overall, the convention would define older people's rights as human rights and demonstrate that the abrogation of human rights is not acceptable. It would stipulate positive obligations on nations to realize equality and the enjoyment of rights by older people. The treaty would considerably expand the concept of human rights protection for older people, since it would not be only about refraining from doing harm or placing negative obligations on the participating states, but would also lay down norms in order to assist older people to attain a status comparable with that of the rest of the population. To achieve these goals, national governments would be required to ensure that the rights set forth in the convention were reflected in their national legislation. (p. 1143).

Along similar lines, the UN Expert Group (2009) concluded:

> A convention on the rights of older persons would add additional weight in furthering, deepening and more precisely defining the rights of older persons. A convention would create obligatory and binding international law. Similar to the adoption of various other human rights instruments, member states would undertake a threefold commitment when adopting such a convention: to respect, to protect and to fulfill the rights enshrined in the relevant text (p. 18).

Despite the clarity of these statements, there is still no strong consensus for a convention for the rights of older persons in the international community. As described by Doron and Apter (2010a, b), there are also voices and arguments against such a convention. Opponents rely on evidence from past experience which indicates that human rights conventions—in and by themselves—do not make a difference to the lives of the disadvantaged. Not only are they not implemented in practice, but in some cases the political compromises that are part of the international bargaining process result in an infringement of material rights. For example, great criticism has been directed at the decision by CEDAW's drafters to ignore honor killings due to political pressure, focusing only on domestic violence (Doron and Apter 2010b). Hence, opponents argue, it is better for older persons to work to improve and implement existing soft law instruments in the field of aging.

While these critiques are important and relevant, Doron and Apter conclude in their article that the counter arguments outweigh them. A convention on the human rights of older people would establish touchstones for legal protection of the rights of older people which are routinely violated. A rising tide lifts all boats, and a convention would set standards for the world. An important outcome of a convention would be the creation of an initial "floor" of bare-bones protections of older persons: a legal "floor" which does not exist today. Implementation of a future convention would include enforcement mechanisms such as a reporting system, ongoing monitoring, and a procedure for individual complaints.

An international convention has impact beyond the signatory countries and beyond its narrow legal impact. Past experiences prove that even countries which are not party to an agreement will look to existing conventions as bench-marks for their legal systems. On the symbolic and education fronts, binding international human rights conventions have proven to be an important tool of political discourse and rhetoric for non-governmental organizations (NGOs) and civil society to change social attitudes towards disadvantaged and excluded social groups.

This was well summarized in 2009 by the (UN Secretary General 2009) regarding the potential importance of a future convention for the rights of older persons:

> A convention would clarify and consolidate existing international norms with respect to the rights of older persons, and it would encourage a more equitable allocation of needed resources for older persons. A convention would clarify the specific obligations of States in order to ensure the full enjoyment of recognized human rights of older persons. It would also empower older persons and provide the framework for national legislation. Moreover, it would provide older persons greater visibility and recognition nationally and internationally as well as the basis for advocacy, public awareness and education on the rights of older persons (p. 16).

Rights protection for older persons in the US and Canada would be far ahead of the provisions of any consensus for a UN convention (at least in its non-discrimination elements). Thanks in part to our system of civil and political rights, older people in North America are important voters, wielding a political force that would be unheard-of in many other countries. Right now, older persons in North America have not rallied behind a convention. However, once a movement to secure the human rights of older people takes hold, it may be hard to stop. That is because where national movements toward equality and non-discrimination go, international movements follow. Conversely, where international conventions lead, countries can follow too.

This would be the more likely scenario for North American countries. US participation in a convention would be an unlikely event in the initial stages, given its historical standoffishness to most other human rights treaties. If the US agreed to participate, signing an international human rights treaty is the first step signaling readiness to implement a treaty, but it is by no means the last. An inter-agency review process that informs consideration by the Senate follows signing. A review would look at existing federal laws, such as the Older Americans Act, as well as state laws. Ratification of any treaty by the United States requires a 'Resolution of Ratification,' that can be achieved only by the support of two-thirds of the Senate. The US' accession in international human rights conventions is a controversial political undertaking. However, in all countries, actual domestic implementation of a convention on the human rights of older persons would be heavily dependent on sub-national dynamics (Hathaway 2005).

Making passage of a human rights convention for the elderly more likely in the US and Canada, is the high participation of the elderly in elections. For example, according to the US Census Bureau, over 70% of voters between the ages of 65 and 74 voted in the 2008 elections. By comparison, only slightly more than 50% of people aged 25–45 voted. Older people tend to vote in much higher numbers than any other age group (US Census Bureau 2008). They claim their political power in a way that many groups facing disadvantages—including groups of older people—cannot do in other parts of the world.

On the international human rights sphere, both on the global and regional levels, the employment rights of older workers have been a blind spot in the field of the rights of older persons. "Hard" international law makes little reference to these issues; it tends to focus on social security, which is an unlikely scenario for most older people. "Soft" international law has no binding effect; and regional instruments (except for those in the European Union) also have failed to directly address these rights in an effective manner. A new international and binding convention, specifically addressing the rights of older persons, might make a difference. To be effective, a future convention must include age as a prohibited ground for discrimination and must state that ageism is illegal in all its manifestations. It should recognize the rights of older workers by including: heightened protections against age discrimination; recognition of the contribution of older workers; heightened job protection to prevent older people from being pushed out of the workplace through harassment; the right to ongoing education and on-work training to allow older workers to adapt to changing technologies and working environments; and flexible and part-time working rights to accommodate sickness or disability.

On an individual level, work and civic engagement can increase health, well-being, and a sense of self-worth in older adults. These traits translate into the human rights concepts of agency; the right to participation in society; the right to health; and the right to be free of discrimination.

On a societal level, the engagement of older persons enriches society. Older persons are not just the recipients of care. They mentor colleagues in the workplace. They are repositories of cultural and institutional knowledge. They have experience that, despite the rapid pace of societal change, is relevant and valuable to society as a whole.

A prominent law professor once wrote, "It is easy to determine whether a state has ratified a treaty; it is much more difficult to evaluate whether it is complying with it" (Hathaway 2005, p. 507). But a convention would do much more than provide another way for countries to support rights A convention would contribute to compliance. It would pave the way for societal change and improve the working lives of older people, as recognition of their rights becomes more routinely protected and commonplace, on equal footing with other groups. This would go a long way toward creating "*A Society for all Ages.*"

References

African Union. (1986). African charter on human and people's rights. *Rights of Older Persons*, p. 161.

APAA (n.d.). *The Arab plan of action on ageing to the year 2012*. http://www.escwa.un.org/popin/docs/Arabplanofactiononageing2012Eng.pdf

Arab Charter on Human Rights (1994, September 15). *Reprinted in Human rights Law Journal*, 18, p. 151. http:www1.umn.edu/humanrts/instree/arabcharter.html

Barrientos, A., Gorman, M., & Heslop, A. (2003). Old age poverty in developing countries: Contributions and dependence in later life. *World Development, 31*(3), 555.

Boyle, A. E. (1999). Some reflections on the relationship of treaties and soft law. *International and Comparative Law quarterly, 48*, 901–902.

Charter of Fundamental Rights of the European Union. (2000). Online: http://www.europarl.europa.eu/charter/pdf/text_en.pdf

Doron, I. (2005). From national to international elder law. *International Journal on Aging, Law and Policy, 1*, 45–72.

Doron, I., & Apter, I. (2010a). International rights of older persons: What difference would an international convention make on the lives of older persons? *Marquette Elder's Advisor Law Review, 11*(2), 367–385.

Doron, I., & Apter, I. (2010b). The debate around the need for an international convention on the rights of older person. *The Gerontologist, 50*(5), 586–593.

Doron, I., & Mewhinney, K. (Eds.). (2007). *The rights of older persons: A collection of international documents* (Universal Declaration of Human rights 1948, pp. 443–444). Israel: International Federation of Ageing.

European Convention for the Protection of Human Rights and Fundamental Freedoms. (1950). Online:http://www.echr.coe.int/NR/rdonlyres/D5CC24A7-DC13-4318-B457-5C9014916D7A/0/CONVENTION_ENG_WEB.pdf

Field, S., & Sappideen, C. (2009). Anti-discrimination – Some observations from down under, the Australian experience on age discrimination. *Journal of International Aging, Law and Policy, 3*, 169.

Gold, J. (1983). Strengthening the soft international law of exchange arrangements. *American Journal of International Law, 77*, 443.

Hathaway, O. (2005). Between power and principle: An integrated theory of international law. *University of Chicago Law Review, 72*, 469.

Human Rights Council (2011, July 4). *Thematic study on the realization of the right to health of older persons by the special Rapporteur on the right of everyone to the enjoyment of the highest attainable standard of physical and mental health.* Anand Grover, A/HRC/18/37

Kanter, A. S. (2009). The convention on the rights of people with disabilities and its implications for the rights of elderly people under international law. *Georgia State University Law Review, 25*, 873.

International Covenant on Civil and Political Rights (ICCPR). (1966). Online: http://www2.ohchr.org/english/law/ccpr.htm

Kye, T. S. (2009). Divided by a common language: Why British adoption of the American anti-discrimination model did not lead to an identical approach to age discrimination law. *Journal of International Ageing, Law and Policy, 17.*

Megret, F. (2011). The human rights of older persons: A growing challenge. *Human rights Law Review, 11*, 37.

Office of the High Commission for Human Rights. (2010). *Current status of the social situation, wellbeing, participation in development and rights of older persons worldwide*. Geneva: United Nations.

Pilapil, L. (2011, May 6). Supreme court panel ends Muhammad Yunus leadership of Grameen Bank. *The Development Newsletter.*

Rodriguez-Pinzon, D., & Martin, C. (2003). The international human rights status of elderly persons. *American University of International Law Review, 18*, 915–1008.

Tang, K., & Lee, J. J. (2006). Global social justice for the older people: The case for an international convention on the rights of the elderly. *British Journal of Social Work, 37*(7), 1135–1150.

US Census Bureau (2008, November). *Reported voting and registration, by age, sex and educational attainment.* http://www.census.gov/hhes/www/socdemo/voting/publications.p20/2008/tables.html

UN (1998). *International plan of action on ageing and United Nations principles for older persons*. New York: UN Department of Public Information (DPI/932/Rev.1-98-24545-September 1998-40M).

UN (2003). *Madrid International Plan of Action on Ageing 2002: Second world assembly on ageing*. New York: United Nations Department of Public Information (DP/2271-February 2003-20M). http:www.un.org/esa/socdev/ageing/madrid_intlplanaction.html

UN Department of Economic and Social Affairs. (2007). *World economic and social survey*. New York: United Nations.

UN Department of Economic and Social Affairs (2008). *The Madrid International Plan of Action on Ageing guiding framework and toolkit for practitioners and policy makers*. New York: United Nations. http://www.un.org/ageing/documents/building_natl_capacity/guiding.pdf

UN Department of Economic and Social Affairs. (2009). *World population ageing*. New York: United Nations.

UN Expert Group. (2009). *The report of the expert group meeting – Rights of older persons*. Bonn: United Nations.

UN Secretary General. (2009). *Follow-up to the second world assembly on ageing, sixty-fourth session*. New York: United Nations.

Venne, R. (n.d.). *Mainstreaming the concern of older persons into the social development agenda*. http/www.un.org/ageing/ducuments/mainstreaming/positionpaper.pdf

Looking Ahead

Patricia Brownell and James J. Kelly

In this book the authors have presented an integrated framework of ageism and abuse of older persons in the workplace. This framework is based on policy and practice guidelines proposed in the 2002 United Nations-sponsored Madrid International Plan of Action on Aging (MIPAA), as well as a complex of variables from the literatures on socio-demographic trends, conceptual definitions of ageism, elder abuse, workplace abuse and bullying, age discrimination, and the co-occurrence of aging and abuse. The contributing authors of this book have applied this framework to an examination of the North American workplace.

The task of integrating multiple frameworks into a single new framework is an ambitious one, but one which our authors have accomplished handily. Their work in this collection provides a firm model for a more comprehensive international treatment of the topic, a task that awaits future scholars and policy analysts. The authors in this collection are international experts in the field of elder abuse. They are affiliated with the International Network for the Prevention of Elder Abuse (INPEA), or are engaged in research and practice related to the workplace. Each author was willing to stretch conceptually to apply international principles of aging and abuse to a fresh examination of an old problem: ageism in the workplace.

Historical Background

This book developed at a unique time in the history of the aging movement worldwide. Following the promulgation of the 2002 MIPAA, the conceptualization of challenges and opportunities in aging is undergoing a reformulation. The paternalistic notion of old age as a time of frailty, dependency and need is being rebalanced with more realistic notions of an ever increasing number of people moving into old age with vigor, health, purpose, education and skills, and a desire to remain productive in the workplace well beyond traditional retirement age. This change has spurred creation of an integrated rights-based aging agenda (MIPAA) that challenges nations

P. Brownell and J.J. Kelly (eds.), *Ageism and Mistreatment of Older Workers: Current Reality, Future Solutions*, DOI 10.1007/978-94-007-5521-5,

to develop an action plan to open up opportunities for older adults so they can participate in society to their full capacity. In addition, older adult workers are beginning to raise their voices about workplace discrimination, abuse and mistreatment (see Chap. 3 in this volume). Media coverage of ageism in the workplace is increasing and social welfare scholars are beginning to re-examine the institution of retirement using an ageism lens.

The present volume focuses on societal barriers in the environment that an international movement by and for older adults can begin to change. This movement is already underway, and the 2002 MIPAA, also known as The Madrid Plan, represents a guide to its direction. The Madrid Plan, as described by Caldera in Chap. 1, embodies the hope for broad social change in making the world a better place in which to grow old with dignity and free of abuse. The foundational document delineating human rights and ageing is known as the United Nations Principles for Older Persons (United Nations 1991).

Personal Resilience and Workplace Barriers

In the book, Resilience in Aging, edited by Resnick et al. (2011), contributing authors focused on older adults' personal resilience in aging as a factor that can buffer negative workplace behaviors, conditions, and policies. The Editors of this collection recognize that not only do older adults need to be resilient, but also the existing attitudinal, structural, social and legal barriers to age-friendly workplaces must be challenged. Elder abuse can occur, even to older adults who are not frail or dependent. And research has shown that even the most resilient older adult can wilt in the face of abuse, bullying and mistreatment, whether in institutional settings, in the home, or in the workplace. Both strengthening environmental protections and promoting personal resilience in the face of abusive workplace practices and policies remain essential to ensuring opportunities for older adults to be fully productive in society.

Litigation and Legislation

Recently the Editors have seen a number of age discrimination cases go to the courts, with mixed results; in a litigious society this is likely to continue. In the United States, for example, an older worker lost an age discrimination lawsuit in the U. S. Supreme Court (Gross v. FBL), making it more difficult legally to prove age discrimination by an employer, even if it involves pain and suffering (Noguchi 2012). However, more recently, legislators introduced a bill to restore older workers' rights by reversing the Supreme Court's age-biased ruling (Editorial 2012).

Although the Editors of this volume welcome such efforts, we take note of Spencer's caution in Chap. 9 against unrealistic optimism that litigation and legislation alone will reduce the problem. Addressing ageism and abuse of older people in the workplace will take a combination of individual, organizational and national interventions from the bottom up and the top down.

A Look to the Future

Researchers and policy analysts have made progress in this field, but definitional terms remain problematic in the design of research studies and legislative and workplace policies. While elder abuse researchers have brought some clarity to definitions in North America, others continue to struggle at the international level (Podnieks et al. 2010).

As the collection of chapters in this edited book suggests, there is the potential for this area of research—ageism and abuse in the workplace—to move in a number of directions. However, the Editors believe that for academics, practitioners and policy makers alike, there is value to shared understanding and insight into this complex and multifaceted social problem. The Editors hope this book promotes conceptual development and debate by establishing common ground across disciplines to address the issues raised here. It is an objective of this book to contribute to bridging the knowledge gap between two conceptual frameworks: that of ageism in the workplace and that of elder abuse.

Having established that ageism is a social issue of growing importance and having identified the role that abuse and bullying of older people can play in the workplace, the Editors hope that more is done at the individual, organizational and societal levels to remedy them. There is a need for more workplace policies, programs and laws addressing these problems from an organizational perspective. But there is also a need for more attention to outcome research and for broadening the context in which elder workplace abuse is examined. The Editors have provided one framework for doing so, by asking contributors to discuss the issues of ageism and abuse in the workplace using an international framework. In the future, research on the impact of ageism and abuse in the workplace from gender, race, sexual orientation and ethnic perspectives can contribute to our understanding as well.

Progress in this field will be enhanced by research findings using the voices of older adults themselves; media coverage to promote positive images of aging; policy makers' understanding of trends in work and retirement policy; examination of innovations in workplace restructuring; changes in human resource policies and practices; new models for promoting resiliency at the individual level; and changes in labor laws and policies to better protect and support older adults. The scholarship presented by each author in this book contributes to this effort.

In the final chapter of this volume, the authors argue that the international movement for the human rights of older adults represents the next great frontier for breaking down barriers to active and productive aging. As the Editors write this in April 2012, the UN is conducting a 10 year review of progress nations have made in implementation of MIPAA recommendations. This process involves examining laws, policies, practices and research undertaken since the MIPAA's passage in 2002. The review, conducted from the "bottom up" (United Nations 2006), is supported by non-governmental organizations with consultative status like the International Network for the Prevention of Elder Abuse (INPEA) and the International Federation of Social Work (IFSW).

Will the next milestone in the human rights movement for older adults be a binding human rights instrument like a convention, as discussed in the final chapter of this volume? All members of society and all nations have a stake in the outcome.

References

Editorial (2012, March 30). Combating age discrimination. *New York Times*, p. A24.

Noguchi, Y. (2012, February 12). *Age discrimination suits jump, but wins are elusive*. Retrieved April 8, 2012, from http://www.npr.org/2012/02/16/146925208/age- discrimination-suits-jump-but-wins-are-elusive

Podnieks, E., Anetzberger, G. J., Wilson, S. J., Teaster, P. B., & Wangmo, T. (2010). Worldview environmental scan on elder abuse. *Journal of Elder Abuse & Neglect, 22*(1–2), 164–179.

Resnick, B., Gwyther, L. P., & Roberto, K. (Eds.). (2011). *Resilience in aging: Concepts, research, and outcomes*. New York: Springer.

United Nations. (1991). *United Nations principles for older persons: adopted by General Assembly resolution 46/91 of 16 December 1991*. Retrieved on April 8, 2012, from http://www2.ohchr.org/english/pdf/olderpersons.pdf

United Nations. (2006). *Guidelines for review and appraisal of the Madrid International Plan of Action on Ageing: Bottom-up participatory approach*. New York: Department of Economic & Social Affairs.

Index

P. Brownell and J.J. Kelly (eds.), *Ageism and Mistreatment of Older Workers: Current Reality, Future Solutions*, DOI 10.1007/978-94-007-5521-5,

CPSIA information can be obtained at www.ICGtesting.com
Printed in the USA
LVOW101215240213

321458LV00009B/298/P

9 789400 755208